G200 Youth Forum 2016 Conference

Garmisch-Partenkirchen, Germany
6 - 10 April 2016

ISBN: 978-1-5108-3254-1

Printed from e-media with permission by:

Curran Associates, Inc.
57 Morehouse Lane
Red Hook, NY 12571

Printed by Curran Associates, Inc. (2017)

G200 Association
14 rue du Rhone
Geneva 1204, Switzerland

Phone: +41-22-518-28-88

g20youth@g200youthforum.org

Additional copies of this publication are available from:

Curran Associates, Inc.
57 Morehouse Lane
Red Hook, NY 12571 USA
Phone: 845-758-0400
Fax: 845-758-2633
Email: curran@proceedings.com
Web: www.proceedings.com

G200
ASSOCIATION

2016 (4)

CONFERENCE PROCEEDINGS

G200 Youth Forum 2016 Conference
April 6-10, Garmisch-Partenkirchen, Germany

Introduction

The G200 Youth Forum (Garmisch-Partenkirchen, Germany, 6-10 April 2016) was one of the largest international events organized for young leaders from around 200 countries in 2016, and over 200 young leaders, students and academics, representatives of the business world and governments were participating in it.

This was the 11th year of the Forum. Previous events were held in:
2006 – G8 Youth Summit – Russia (Saint-Petersburg);
2007 – G8 Youth Summit – Germany (Berlin);
2008 – G8 Youth Summit – Japan (Tokyo);
2009 – G8 Youth Summit – Italy (Milano);
2010 – G20 Youth Summit – Canada (Vancouver);
2011 – G20 Youth Summit – France (Paris);
2012 – G20 Youth Summit – USA (Washington D.C.);
2013 – G20 Youth Forum – Russia (Saint-Petersburg):
2014 – G20 Youth Forum – Germany (Garmisch-Partenkirchen, Bavaria);
2015 – G200 Youth Forum – Germany (Garmisch-Partenkirchen, Bavaria); and
2016 – G200 Youth Forum – Germany (Garmisch-Partenkirchen, Bavaria).

The G200 Youth Forum 2016 had 3 main events which run in tandem with each other:
G200 Youth Summit – Communiqué
Conference – Publication of academic articles in the Conference Proceedings
International Young Parliamentarians' Debate – Joint Statement

Conference 2016 was an international Conference for representatives from the best Universities in the world who were experts in international relations, law, economics, finance, technology, medicine, education and humanities. During the Conference around 300 participants discussed global problems which were on the agenda of the academic communities of different countries in the format of 8 round tables working in parallel.

1. Economics and Finance
2. Law and Human Rights
3. World Politics and International relations
4. Social Affairs and Medicine
5. Ecology, Environment, and Energy
6. Design, Technology, and Innovations
7. Education and Youth
8. Humanities: History, Philosophy, Linguistics, Arts and Journalism

In total: more than 4700 attendants in 11 years from around 200 countries.

The participants of the Conference were representing about 40 best Universities of their countries:

Al Yamamah University
Bifrost University
Centennial College
Deakin University
Erasmus University Rotterdam
European University of Barcelona
European University, International Business School
Gadjah Mada University
Gambia Technical and Training Institute
Griffith University
HEC Montreal
Higher School of Economics
Hokkaido University
Hong Kong University of Science and Technology
Instituto Universitário de Lisboa
ITMO University
Keele University
Korea University
Michigan State University
Nankai University
North-Eastern Federal University
Osaka University
Peking University
Punjab Technical University
Tamkang University
Tianjin University of Finance and Economics
Tokyo Institute of Technology
Tokyo University of Foreign Studeies
The Ohio State University
Universidad Autonoma Metropolitana
Universitas Airlangga
Université de Lausanne
Universiteit Leiden
University of Aberdeen
University of Central Lancashire
University of East Anglia
University of Granada
University of Hertfordshire
University of Melbourne
University of Pécs
University of Pretoria
University of Queensland
University of St. Andrews
University of Winchester

Reviewer Acknowledgement for the Conference Proceedings 2016

G200 Association would like to thank all our reviewers who have contributed to the Conference Proceedings 2016.
Dr. Zhang Yanling, Professor, Deputy Director of Tourism Department, Business School, Tianjin University of Finance and Economics, Tianjin, China.
Mr. Asaad Al Asaad, Director of Student Activity Administration, Al Yamamah University, Riyadh, KSA.

In the Conference Proceedings 2016 the spelling and punctuation of the authors are kept. The opinions expressed in the articles are those of the authors and do not necessarily reflect the position of the G200 Association.

Content

Economics and Finance

Session I: Investing and Financing

The Role of Innovation for Entrepreneurial Start-up and Firm Growth

Xiangyi Feng
Glenmere Technology Ltd, United Kingdom

EXECUTIVE SUMMARY

With the development of the economy and under the help of internet, entrepreneurs seek to start up their own business. What is more, the society has also provided assistant to the entrepreneurs. So it is easier to start up nowadays. However, if a person wishes to achieve success, he should also own some characteristics that are beneficial to start-up. In such condition, this report aims at discussing the innovation for start-up and firm growth. Innovation will influence the identification of entrepreneurial opportunities, start up and firm growth. According to the statement, this report also put forward some suggestions that are beneficial to the development of companies.

1. INTRODUCTION

With the rapid development of the economy, people have higher pursuit of the wealth and personal value. What is more, under the help of the network, people can receive the information from everywhere, so if someone decides to undertake business, it will become easier to them. However, entrepreneurship seeks to create something new, such as new markets, new production processes, and new ways of organizing existing technologies. What is more, it also emerges out of the intersection of inspiration and activation and is influenced by many factors. So it is crucial to pay attention to the influence of innovation on entrepreneurship. First, the report will briefly introduce the definition of innovation and entrepreneurial start-up (Robert & Baron 2013). Meanwhile, it will also summarize the factors that are affecting innovation and entrepreneurial start-up. Secondly, while innovation will affect entrepreneur, it will have different impacts at different stages. At the early stage of entrepreneurship, in order to start a business, the people should also identify entrepreneurial opportunities and innovation and creativity influence identifying entrepreneurial opportunities (Zahra & Covin 2013). Then the report will demonstrate the influence of innovation on entrepreneurial opportunities, entrepreneurship start-up, and firm growth. What is more, it will put forward some recommendations to promote innovation and it consists of four aspects. On one side, it promotes innovation through encouraging and enhancing creativity. On the other side, cultivating the consciousness of innovation and emphasizing entrepreneurship education are also important. Meanwhile, it can improve entrepreneurial management mechanism. Last but not the least; the company can establish supporting system of innovation and entrepreneurship. At last, it will have a conclusion about the whole report.

2. METHOD

In order to analyse the relationship between innovations an entrepreneurship, the report analyses the relation between them through case study. Through analysis of 21 entrepreneurs and some famous companies, it is better to have a good command of the relationship between innovation and entrepreneurship.

3. ANALYSIS OF THE LITERATURE

3.1 Innovations and Entrepreneurial Start-up

Innovation is that involving the conversion of new knowledge into a new thing, service and the putting this new thing, service into actual use. From reviewing the literature, it discovers that many scholars have studied the influencing factors of innovation (Arthur, Bell & Edwards 2013). Scott and Bruce consider that the leaders, team, and individual characteristics affect innovation. What is more, Wolfe put forward four factors that influencing individual innovation, these factors are organizational structure, innovation environment, culture environment, and individual characteristics. Ong Wan (Baron 2014) in Singapore also infers that the lever of employees in organization, internal reaction and communication between departments, leadership, supporting coming from leaders and the challenges of the job impact on innovation. Besides, other scholars have also made researches on innovation. According to the above statement, it can indicate that there are lots of factors that influencing innovation, so if a person decides to start a business, he should take many factors into account.

As a field of business, entrepreneurship seeks to recognize how specific individuals find chances by using various means to create new markets, new production processes, new ways of organizing technologies and producing a wide range of effects

(Tomasz Krzys & Michael 2012). With the rapid development of the economy, more and more experts pay attention to start-up, and there is a tendency that the students who graduate from schools will become the main force of entrepreneurship. There are many factors influencing entrepreneurship; the factors include environment, technology and skills and so on.

3.2 The Influence of Innovation on Entrepreneurship at Different Stages

Identifying entrepreneurial opportunities is the start of entrepreneurship and determines whether entrepreneurship is successful or not. To the entrepreneurs, they should own the ability to identify the entrepreneurial opportunities which are mostly affected by individual characteristics (Savvas, Feng, Henry & Michael 2013). According to the investigation of 21 entrepreneurs, it has discovered that most of the entrepreneurs have the belief to pursue progress, innovation and wealth and strong insight to grasp the information of the market. Shane indicates that entrepreneurs could innovate new products or service, new processes, new market and new material by taking advantage of technology, politics and the change of market demand (Lowell & Busenitz 2013). When a person decides to identify entrepreneurial opportunities, first he should hunt for opportunities according to the targeted search of the whole economic system. Secondly, through the judgment of market and analysis of the innovation, he identifies entrepreneurial opportunities. Last but not the least, after identifying entrepreneurial opportunities, it is critical to evaluate the opportunities. It should evaluate every indicator of entrepreneurial opportunities, such as the evaluation of finance and the structure of the team. Through the evaluation of entrepreneurial opportunities, the entrepreneurs can determine whether to start up.

As one kind of innovations, technology innovation has made a great contribution to the entrepreneurship. From the surveys of 21 entrepreneurs, it discovers that 23% of the entrepreneur's start up their business through technology innovation. They turn the difficult information and knowledge into simple and understandable one by taking advantage of computers and Internet. Thus, everyone can be aware of the changes quickly and realize the benefits of network. Through Internet, they can integrate the resources what they need. What is more, through network, it is more convenient to start up.

The cores of entrepreneurship are integration of resources and teams and the creativity and innovation of entrepreneurs, and entrepreneurial teams determine if they will achieve the resources. Timmons suggests that the entrepreneurship is a dynamical process, and it should match entrepreneurial opportunities with an entrepreneurial team and entrepreneurial resources. Salman considers that there are four key factors that are affecting entrepreneurship. If a company can focus on innovation at the stage of start-up, the company can achieve success (Christian, Pierre & Andre 2014). For example, at an early stage, Ma due combines B2B with Internet and establishes Alabama quickly. Up to now, Alabama has achieved great success. In addition, there are many other have achieved success through innovation, such as Microsoft. After identifying entrepreneurial opportunities, Bill Gates establishes Microsoft under the help with Internet (Natasha & Colm 2011).

Innovation has a close relation with firm growth. The firm growth is a cycle process conforming to the life cycle theory. The firms would go through the period of start-up, growth, maturity and decline. At the early stage of entrepreneurship, innovation is the theme of the companies. In growth period, system design, industry selection, and industry diversification are the key points of firm growth. What is more, they mostly include structure innovation and technology innovation (Shane & Venkataraman 2013). After the forehead innovation and accumulation, the companies enter the peak state and achieve competitive advantages on production technology, quality and sales channels to resist market risk. Innovation will impact on many aspects of companies in the period of growth.

Innovation can change the pattern of world business. Some emerging businesses take advantage of innovation and rapidly become the leader who can formulate the rules of the game in the industry (Zahra & Dess 2012). For instance, Bill Gates and his company have won the dominance of computer software industry by developing microcomputer and computer operating system which have changed people's lifestyle and style of work and were popular to the consumers. Besides, innovation would also lead to the change of operation mode of the company. In order to create more profits, some established companies enhance their abilities of innovation. Wanxiang Company is built in 1969 and has survived for 40 years in the heated competition, because Wanxiang Company insists on its way of innovation and put the improvement of self-innovation in the first place, so it has fulfilled steady economic growth for 40 years (Leonidas, Zampetakis & Vekini 2011). Since Lafley acts as CEO of Procter & Gamble, he put forward putting innovation into all activities of P&G,

and set an example to others, which has saved P&G in the decline and creating huge profits.

What is more, innovation can help companies enter the market faster and deeper. For example, Nokia adds some functions according to the unique needs of rural consumers as a result of increasing two million customers in India. Besides, with the aid of innovative business model, Dell has made a huge success by direct sales and order production. Innovation can not only fulfill sustainable growth, but also act as the catalyst of organizational strategy, culture and leadership. If a company wishes to become bigger and stronger to get long-term development, it should build up innovation environment and culture. P &G has carried out innovation culture under the leadership of Lafley and has got great growth in the past eight years. What is more, P& G has established it's branched all around the world and creates huge profits.

According to the above statement of the role of innovation for entrepreneurship, the report also put forward framework of entrepreneurship process basing on creativity and innovation of entrepreneurs. That is to say, through participating in every stage of entrepreneurship, the entrepreneurs receive creativity and innovation to identify entrepreneurial opportunities, construct new company and achieve firm growth (Stern & Jonathan 2011).

3.3 Recommendations

Innovative companies also exist some limitations and constraints. However, innovation is an inexhaustible power. In comparison with the competitors who are larger and stronger, the company should break through these limitations to achieve more market share to fulfill success (Lorraine& Roy 2013). To the innovative entrepreneurship, the spirit of innovation should be taught to every department and employee. Through training, the responsibility and creativity of employees will be enhanced and improved. What is more, it is beneficial to put innovation into the business activities, and innovation could be the foundation of companies. In order to improve innovation, the report also put forward some suggestions.

Creativity is the process that transforming the ideas of the organizational members' ideas into a useful product, service or working manners. What is more, creativity is the basis and source of innovation. It is encouraged to have more creativity that is deserved to protect (Jorrow & Yu 2011). What is more, it is important to respect other people's comprehension and open mindedness on knowledge. At the same time, the companies should also encourage doubts and imaginations to the problem of the employees to improve the innovation of the employees. From the above statement of examples, the successful entrepreneurs achieve their success through finding creativity. Ma due establishes Alibaba through identifying the entrepreneurial opportunities of Internet as well as Microsoft has created huge profits by innovation and creativity.

First, the society should build up environment and atmosphere which support encouragement, innovation, risk-taking and tolerance of failure. Innovation atmosphere is critical to the entrepreneurship. If the environment is stable and favourable, identifying entrepreneurial opportunities will become easy (Ans & Bart 2014). Favourable environment will improve the entrepreneurs' ability of start-up and entrepreneurs will identify entrepreneurial opportunities easier. About the favourable atmosphere in the society, people will be more willing to talk about entrepreneurship and take the chances of entrepreneurship.

Secondly, the school should also pay attention to the subject status of students and stimulate intrinsic motivation of students. What is more, about solving skills of students, it is necessary to cultivate their sense of responsibilities and initiative. Nowadays, many companies recruit students who are graduating from college. What is more, after graduation, the students are willing to start their business. Therefore, before students become entrepreneurs, the students should be well trained (Bechard & Toulouse 2012). The entrepreneurs should enhance their ability of reaction, the perception of risk and confidence and so on (Luciano, Sara & Zhan 2013). According to the survey of 21 entrepreneurs, it can derive that most of them own these abilities that have made a great contribution to the development of their companies (Jorunn, Eina & Madsen 2011). Ma due, Bill Gates and other entrepreneurs own these abilities and will develop these abilities in the midst of entrepreneurial career.

First, the company should implement innovation of management and change traditional thoughts of management. Through the cross operation among companies, the companies should provide free space to carry out business with risk. Secondly, the companies should also implement multiple incentives and improve the mechanism of entrepreneurial drive. In order to make entrepreneurship with strong vitality and motivation, it is necessary to improve the mechanism of entrepreneurial activities. To the entrepreneurs, they can not only be motivated by concept, targets, and interests, but also by properties and rights which are innovative ways to motivate the entrepreneurs (Sahlman & Stevenson 2012). At the same time, Haier has changed its management mechanism. Haier combines employees with the customers to achieve a win-win situation. In this pattern, the employees take salary according to the creation of value of the customers rather than the completion of the tasks (Jean 2011). This innovation of management mechanism is to adapt to the demand of Internet.

It should also enhance the supporting system of high-tech industry. What is more, cultivating favourable talent environment is also important to the development entrepreneurship. If the supporting system has been well established, the entrepreneurs will be easy to start up new business, and achieve great growth. Wanxiang has established better supporting mechanism to guarantee that the employees can fully imagine and create new ideas during work without punishment. At the same time, through the establishment of supporting system. The economy of Wanxiang has improved.

4. CONCLUSION

In this era, the competition between companies is very fierce, so if the companies wish to survive in the heated competition, they should create the ways which are different from other companies and innovation is beneficial to companies. Innovation is important to the companies and someone who wants to open his business. From the analysis of the definition of innovation and entrepreneurship start-up, it can indicate that innovation is different from entrepreneurial start-up and innovation can improve entrepreneurship start-up. Innovation is the guarantee of the development of the company. The creativity and innovation are the most important capital of the company. This report has suggested the influences of innovation on entrepreneur. According to the above statement, entrepreneurship has three stages and innovation affects all of them. What is more, the report also put forward some recommendations. For one thing, if a company wants to improve innovation, it should take measures to encourage and enhance the creativity. For another thing, it is necessary to cultivate the consciousness of innovation and emphasizing entrepreneurship education. Moreover, through innovation of entrepreneurial management mechanism, the company will take responsibility, power, and profits into consideration when starting up. Meanwhile, supporting system of innovation and entrepreneurship are also critical to the entrepreneurial start-up (Pedro, Alexandre & Clara 2012). If the supporting system is well established, the company will grow up better in the heated competition. According to the recommendations, the report wishes that the recommendations will be beneficial to the future development of the new company.

REFERENCES

- Ans, H. and Bart, C. (2014). How and Why do Research-Based Start-Ups Differ at Founding? A Resource-Based Configurational Perspective. *The Journal of Technology Transfer, 7(4),* 3-4.
- Arthur, W., Bell, S., and Edwards, B. (2013). A Longitudinal Examination of The Comparative Criterion-related Validity of Additive and Referent-shift Consensus Operationalizations of Team Efficacy. *Organizational Research Methods, 10(12),* 35–58.
- Baron, R. (2014) 'Psychological perspective on entrepreneurship: Cognitive and social factors in entrepreneurs' success. *American Psychologist, 5(6),* 12-15.
- Bechard, J. and Toulouse, J. (2012). Validation of a Didactic Modal for the Analysis of Training Objectives in Entrepreneurship. *Journal of Business, 12(2),* 34-35.
- Christian, B., Pierre, N., and Andre, J. (2014). Defining the Field of Research in Entrepreneurship. *Journal of Business Venturing, 8(12),* 15-19.
- Jean, C. (2011). Revitalizing Entrepreneurship: How Visual Symbols are Used in Entrepreneurial Performances. *Journal of Management Studies, 6(5),* 12-15.
- Jorunn, G., Einar, L., and Madsen, O. (2011). The relationship between resources, entrepreneurial orientation and performance in farm-based ventures. *Entrepreneurship & Regional Development, 6(9),* 3-4.
- Jorrow, R. and Yu, F. (2011). Counterparty risk and the pricing of defaultable securities. *Journal of Finance, 56(5),* 1765-1800.
- Leonidas, A., Zampetakis, M., and Vekini, V. (2011). Entrepreneurial orientation, access to financial resources, and product performance in the Greek commercial TV industry. *The Service Industries Journal, 6(12),* 89-92.
- Lorraine, U. and Roy, T. (2013). Postmaterialism influencing total entrepreneurial activity across nations. *Journal of Evolutionary Economics, 7(2),* 156-160.
- Lowell, W. and Busenitz, G. (2013). Entrepreneurship Research in Emergence: Past Trends and Future Directions. *Journal of Management, 12(7),* 46-47.

- Luciano, C., Sara, B., and Zhan, G. (2013). The inception of internationalization of small and medium enterprises: The role of activeness and networks. *Journal of Business Research, 5(6),* 67-69.
- Natasha, E. and Colm, G. (2011). Improvised internationalization in new ventures: The role of prior knowledge and networks. *Entrepreneurship & Regional Development, 15(2),* 7-8.
- Pedro, C., Alexandre, M., and Clara, G. (2012). The performance of entrepreneurial small and medium sized enterprises. *The Service Industries Journal, 15(7),* 78-79.
- Robert, A. and Baron, J. (2013). The role of entrepreneurs in firm-level innovation: Joint effects of positive affect, creativity, and environmental dynamism. *Journal of Business Venturing, 5(1),* 78-80.
- Savvas, P., Feng, Li. Henry, E., and Michael, C. (2013). Entrepreneurial networks: A Triple Helix approach for brokering human and social capital. *Journal of International Entrepreneurship, 3(9),* 78-82.
- Stern, N. and Jonathan, L. (2011). The entrepreneurial perceptions of strategy makers: Constructing an exploratory path in the pursuit of radical growth. *Journal of Business Research, 7(18),* 56-59.
- Sahlman, W. and Stevenson, H. (2012). The entrepreneurial venture. Boston: Harvard Business School Publication.
- Shane, S. and Venkataraman, S. (2013). The promise of entrepreneurship as a field of research. *Academy of Management Review, 25(1),* 102-105.
- Tomasz, O., Krzys, Z., and Michael, G. (2012). Dominant Logic and Entrepreneurial Firms' Performance in a Transition Economy. *Entrepreneurship Theory and Practice, 1,* 16-20.
- Zahra,S.andDess,G.(2012)Entrepreneurship as a Field of Research: Encouraging Dialogue and Debate. *The Academy of Management Journal, 23(4),* 45-47.
- Zahra, S. and Covin, G. (2013). Contextual influences on the corporate entrepreneurship performance relationship: a longitudinal analysis. *Journal of Business, 45(9),* 56-61.

Tadawul Saudi Stock Exchange: Does Quality Investing Matter?

Dalya Mohamed Bahanshal
Student, Al Yamamah University, Saudi Arabia

ABSTRACT

This research has examined the level of awareness of Saudi community about Quality Investing and their interest regarding implementation of a Quality Index in Tadawul. The research consists of primary and secondary data in which primary data was gathered using a survey questionnaire and secondary data has been collected electronically through institutional websites, ProQuest database, and government websites. According to research findings, majority of participants were not aware of Quality Investing, however most have agreed that quality index should be implemented in Tadawul. Moreover, this research proposes a quality index model called Smart Beta Index model for Tadawul to construct its quality index that will be named the TASI-10 Quality Index. In addition the model is based on two factors which include high profitability and low investments and in order to construct the index 50% of companies should achieve a high profitability ratio and 50% of companies should have low investment ratio. The proposed Quality Index will constitute of companies that are listed in Tadawul and a winner of the national quality award which is the King Abdulaziz Quality Award.
Key words: Quality Investing, Smart Beta, Tadawul, Quality

1. INTRODUCTION

Today through globalization and financial integration, companies awareness of implementing quality standards have increased and people started to realize the positive relationship between profitability and quality. For instance, Currier and Chet (2006) believe that many stock market investors started viewing quality as the new differentiation point which is more familiar in the bond market. As quality conscious customers have increased, more organization have started to comply with quality standards in their business operation to compete for national and international quality awards such as the Malcom Baldrige National Quality Award (MBNQA) in the United States and Deming's prize in Japan. Moreover, according to Bloomberg Business News (1995) financial analysts have been realizing that investment in companies winning quality awards such as MBNQA show higher returns than companies without. In addition, those companies that have high quality ranking minimize their losses during economic downturns such and recover gracefully due to their superior quality management in their business processes. Therefore, the term quality investing became more popular and is now a top investment strategy.

Kamdar (2012) defines Quality Investing as a portfolio constructed from high quality business holdings that have the potential to deliver healthy long-term wealth with minimum volatility. In addition Kamdar (2012) defined high quality business are those businesses that are characterized with consistent operating cycle, strong performance on a yearly basis, consumption oriented themes backed by domestic consumption, low capital intensity and leverage, and high dividend payout. Furthermore, because quality has become an important factor for long-term wealth creation and minimum volatility, quality indices such as the MSCI Quality Indices and S&P 500 High Quality Rankings Index has been developed. Quality indices have been highly successful in developed countries and have been spreading in the emerging countries. According to Tadawul website (2014), Saudi Arabia has one of the most efficient stock exchange markets ranking 21st largest market in the world in 2014, yet they still have not developed a quality index. Therefore this paper will examine the Saudi community's awareness on quality investing and proposes a quality index model for Tadawul.

OBJECTIVE OF THE STUDY

This research aimed to evaluate the level of awareness about Quality Investing in Saudi Arabia, and to propose a quality index model for Tadawul. The following objectives include:

1. To understand the level of awareness of Quality Investing among the Saudi community.
2. To propose Tadawul to construct a TASI-10 Quality Index using Smart Beta Index model.

2. TADAWUL STOCK EXCHANGE MARKET PROFILE

2.1 Tadawul Stock Exchange Market Overview

Tadawul is the Saudi Arabia stock exchange that includes 170 listed companies from 15 different sectors and is regulated by the Capital Market Authority (CMA) that regulates and monitors the activities of the market. Earlier in 1984 the capital market was regulated by ministerial committee including the Saudi Arabian Monetary Agency (SAMA) and later in 2003 the CMA was established and became the sole regulator of the Saudi capital market. Moreover, CMA plays a major role in maintaining the fairness and efficiency of the market. Tadawul's vision is to be an integrated financial exchange that contributes in diversifying the Saudi capital markets and compete in a global scale. Furthermore, its mission is to offer superior value to its market participants and stakeholders by providing them with attractive products and services. In 2014 the Saudi Stock Exchange has been awarded as being the 'Best managed financial exchange' in the Middle East. Recording a trading value of 2 trillion in 2014, Tadawul has been recognized as being the most liquid platform in the Middle East and North Africa. Today Tadawul stands as being the 21st largest stock market in the world surpassing Bursa Malaysia, the Mexican Exchange and the Moscow Exchange.

2.2 Tadawul All Share Index (TASI) Overview

TASI (Tadawul All Share Index) is an index developed by the Saudi stock exchange which was first developed in 1984 with a base point 1000 and it was restructured in 2008. Small trades with trade value less than 15,000 S.R are excluded from the index and are found in the VOLTADA index. Moreover, Quality Index is not found in Tadawul and people are not aware of the contribution it will make if it was available. Benefits of Quality Index to the market participants include: stable return in the long-term, decrease in losses during crucial times and is consistent over the time.

3. SASO NATIONAL QUALITY AWARD:

Under the supervision of The Saudi Standards, Metrology and Quality Organization (SASO), King Abdulaziz Quality Award (KAQA) has been recognized as the only national award and has been established by a royal decree No.7/B/18670 on 4th of March, 2000. The award cycle for KAQA is conducted every five years and the winners of the award include:

- Saudi Telecommunications Company
- Saudi Electricity Company
- Savola Packaging Systems Company
- Jubail Petrochemicals
- Advanced Electronics Company

4. RESEARCH METHODOLOGY

This paper consists of primary and secondary data that included both qualitative and quantitative data. The primary data was gathered from a survey questionnaire that aimed to evaluate participant's awareness on quality investing. The survey was randomly distributed to 70 participants and 53 have responded. Moreover, the survey consisted of 9 key question and it included one direct question about quality investing while the rest were indirect question about quality investing that determined the level of awareness in that field.

Secondary data were gathered from reliable articles and reports that were collected from credible sources such as ProQuest database, government websites and other institutional websites.

5. FINDINGS OF SURVEY AND INTERPRETATION

5.1 Survey Results

Table 1. Descriptive Statistics

Gender			
Female		Male	
79.25%		20.75%	
Ages			
18-24	25-34	35-44	45-54
64.15%	28.30%	5.66%	1.89%

The results of the survey have shown that more than half of the respondents' age ranked between 18-24 with 64.15%, followed by 28.3% between 24-35 and minority of around 7% were between 34-54. From these respondents, 79.25% were female and 20.75% were males. The participants were asked whether they heard about Quality Investing, majority of 69.81% have not heard of Quality investing, while 30.19% have heard about Quality Investing.

Moreover, more than half of respondents define Quality in the equity market as 'high stock price shows high quality', while minority of 9.43% define it as 'low stock price show high quality', followed by 35.85% define Quality in equity market as 'high profitability and low investment'. Later on participants were asked what indicates Quality for a company to be listed in a stock exchange market and many respondents said that 'company with high reputation' indicates quality, followed by 26.42% of respondents have said an 'ISO certificate' is an indicator to Quality. In addition, 22.64% of the participants have said that 'high net-worth company' is an indication for Quality while minority of 18.87% have said 'company winning a national or international award such as Deming's price or MBNQA' is indication to Quality.

Table 2. Survey Questionnaire Results

Questions	Results
Have you heard about Quality?	
Yes	30,19 %
No	69,81 %
Define 'quality' in Equity market	
High stock price show high quality	54,72 %
Low stock price show high quality	9,43 %
High profitability and low investments	35,85 %
What is the indicator of quality for a company to be listed in a stock exchange market?	
ISO certificate	26,42 %
Company with high reputation	32,08 %
Company winning a national or international quality award such as Deming's prize or Malco	18,87 %
High net worth company	22,64 %
Where would you invest to minimize losses in bad times and gain long-term creation of wealth?	
I would invest in bond portfolios	18,87 %
I would have my own portfolio investing in different companies	52,83 %
I would invest in specific Industry index	22,64 %
I would invest in Quality Index	5,66 %
Do you know any of the following Quality Index?	
S&P 500 High Quality Ranking Index	28,30 %
MSCI World Quality Index	20,75 %
FTSE Developed Quality	11,32 %
No	58,49 %
Would you like to have quality investing in KSA through a Quality Index in Tadawul?	
Yes	92,45 %
No	7,55 %

Furthermore, to test respondents' knowledge in Quality Investing, they were asked to choose an investment strategy that would minimize their losses in crucial times and deliver long-term wealth, results show that more than half have chosen 'I would have my own portfolio investing in different companies' as an investment strategy. In addition, 22.64% have chosen to invest in a specific industry index, followed by 18.87% chose to invest in bond portfolio, while minority with 5.66% have chosen to invest in a Quality index as an investment strategy. Later on, participants were asked whether they know the given Quality indices with the option to choose more than one and majority of 58.49% have said 'no', followed by 28.3% have said they know 'S&P 500 High Quality Ranking Index', while 20.75% have chosen 'MSCI World Quality Index', and minority of 11.32% have chosen FTSE Developed Quality. Finally participants were asked whether they want to have a Quality Index in Tadawul (Saudi Stock Exchange) and a vast majority (92.45%) of participants have said 'yes' while 7.55% have said 'no'.

5.2 Analysis and Interpretation of Survey Findings

The study show that minority were aware of Quality with 30,19%, however, findings show the respondents that knew about quality investing didn't respond correctly to the indirect questions that were asked to assess the awareness of participants about quality investing. First, the major indicator of a company's quality in a stock exchange market is the national/ international quality award which was only chosen by 18.87% of participants; while high reputation was chosen by 32.08% of participants. Second, participants were given a scenario to choose an investment strategy that will help them to minimize losses in crucial times and gain long-term creation of wealth and only 5.66% of participants have chosen the appropriate strategy to invest in a quality index. Therefore, this shows that there is no correlation and a mismatch between the percentage of participants that knows about Quality Investing (30.19%) and 5.66% of participants that have chosen to invest in quality index.

Figure 1. The Smart Factor Multi-Strategy Index Construction Model

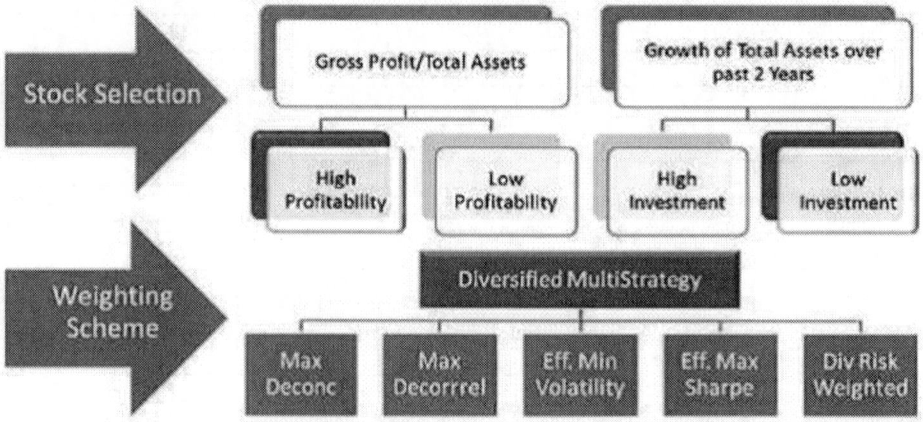

Source: The Smart Factor Multi Strategy Construction Model. Reprinted from The Dimensions of Quality Investing: High Profitability and Low Investment Smart Factor Indices, by Scientific Beta An EDHEC-Risk Institute Venture, retrieved from http://www.scientificbeta.com/#/. Copyright 2015 by ERI scientific Beta

Moreover, it was expected to find correlation between those that heard about Quality investing and their response to the relative scenarios but that was not the case. However, the one time that participant may have some kind of knowledge about Quality Investing was when they were asked to define quality in Equity markets and 35.85% have chosen 'high profitability and low investment' which was the most appropriate for this given scenario. Furthermore, participants were asked if they knew about the given quality indices that included S&P 500 High Quality Ranking Index, MSCI World Quality Index, and FTSE Developed Quality and majority didn't knew about these quality indices. The person that knows about Quality Investing should know about these quality indices because they are not only used as benchmark but are also used in the development and construction of quality indices. Even though Quality Investing was not a familiar topic for most of the participants, however vast majority were interested and agreed to have a Quality Index in Tadawul. Overall, from the survey results we saw a lack of awareness of participants about Quality Investing and high interest on the implementation of Quality Index in Saudi Arabia.

6. PROPOSED QUALITY INDEX MODEL FOR TADAWUL

The proposed quality model for Tadawul is the Smart Beta Index model that is based on the Smart Beta 2.0 approach. This approach constructs index based on the Smart Beta method which involves exposure of both 'beta' and 'smart weighing'. The beta exposes investors to the desired risk while the smart weighting reduces the unrewarded risk through diversification. Construction will involve stock selection and weighting scheme. Stock selection will be based on two factors which is high profitability measured by gross profitability and low investment measured by total asset growth over two years. The two smart factors indices provide a smart way to diversify risk and capture the risk premia. According to Scientific Beta USA Long-term Record (2015), combining both high profitability and low investment leads to high annual returns, low volatility, high Sharpe ratio, and low Max Drawdown in comparison with stand-alone indices. The proposed index for Tadawul will be called the TASI-10 Quality Index that will constitute of ten companies that have the King Abdulaziz Quality Award (KAQA). Moreover, the ratio for high profitability and low investment will be calculated and 50% of high profitability ratio companies and 50% of low investment ratio companies will be selected to construct the index. Once we have selected the stocks, the second step is the smart weighting which involves computing five diversification weighting scheme that includes Maximum Deconcentration, Maximum Decorrelation, Efficient Minimum Volatility, Efficient Maximum Sharpe Ratio and Diversified Risk Weighting. Furthermore, the five weighting schemes results in a diversified multi-strategy index that involves an annual scoring, quarterly stock selection, quarterly rebalancing, daily valuation and quarterly equal allocation of the two factor indices (profitability and investment).

According to the Scientific Beta EDHEC Risk Institute Venture (2015) the Smart Beta also referred as the Scientific Beta outperformed its competitors when a comparison was made between the Smart Beta and World Quality Index competitors based on a 10-year historical data for both US and Developed region. In

Figure 2. Performance Comparison between Smart Beta Index and World Quality Index
Using S&P500 Index as a Benchmark

Dec 2004 – Dec 2014 (10 Years)	S&P 500 Index	Scientific Beta USA			Competitors USA		
		Low Investment Multi-Strategy	High Profitability Multi-Strategy	Custom EW Combination	MSCI USA Quality Index	Russell 1000 Quality (HEFI)	S&P 500 High Quality Rankings
Absolute Analytics							
Ann. Returns	7.65%	10.88%	10.59%	10.75%	9.05%	9.50%	7.79%
Ann. Volatility	20.39%	18.81%	18.75%	18.70%	18.12%	19.34%	20.52%
Sharpe Ratio	0.30	0.50	0.49	0.50	0.42	0.42	0.31
Max Drawdown	55.25%	50.82%	47.58%	48.95%	44.03%	48.61%	57.68%

Source: Performance Comparison using S&P 500 Index as a benchmark. Reprinted from The Dimensions of Quality Investing: High Profitability and Low Investment Smart Factor Indices, by Scientific Beta An EDHEC-Risk Institute Venture, Retrieved from http://www.scientificbeta.com/#/. Copyright 2015 by ERI scientific Beta

the US, S&P 500 Index was used as a broad market index benchmark and results show that the 'Smart Beta' outperformed both benchmark and competitors in terms of annual return, annual volatility, Sharpe ratio, and Max Drawdown. In the Developed region, MSCI World Index was used as a benchmark and the results also showed that the 'Smart Beta' once again outperformed both the benchmark and competitors and for both scenarios competitors were the MSCI USA Quality Index, RUSSEL 1000 Quality (HEFI), and S&P 500 High Quality Rankings.

The major limitation for this model is that from the five companies that have the King Abdulaziz Quality Award (KAQA) only three are listed in Tadawul. Moreover, for the model to be constructed it's required to have a minimum of 10 companies listed in Tadawul with King Abdulaziz Quality Award and from these 10 companies 50% should achieve a high profitability ratio and 50% should achieve low investment ratio. However with the coming years more companies will win the award and meet the requirement and thus the TASI-10 Quality index can be constructed.

7. CONCLUSION

Based on primary and secondary research, the researcher has concluded that majority of the participants are not aware of Quality Investing. In addition the participants that were aware with quality investing were found out that they were not familiar with the topic when different scenarios were given to them about quality investing. Although majorities were not familiar with the topic, however, most have agreed that quality index should be implemented in Tadawul. In Saudi Arabia the only quality award is represented by King Abdual Aziz Quality Award and it has been found that five companies have obtained the quality award. Moreover, the proposed quality index model for Tadawul is the model developed by Scientific Beta EDHEC-Risk Institute Venture called the Smart Beta Index model. Smart Beta is the method used to construct the index that is based

Figure 3. Performance Comparison between Smart Beta Index and World Quality Index

Dec 2004 – Dec 2014 (10 Years)	MSCI World Index	Scientific Beta Developed			Competitors Developed		
		Low Investment Multi-Strategy	High Profitability Multi-Strategy	Custom EW Combination	MSCI World Quality Index	Russell Dev Quality (HEFI)	FTSE Developed Quality
Absolute Analytics							
Ann. Returns	6.59%	9.60%	9.84%	9.73%	8.95%	8.73%	8.29%
Ann. Volatility	17.49%	15.31%	15.39%	15.31%	16.77%	16.90%	16.07%
Sharpe Ratio	0.29	0.53	0.55	0.54	0.45	0.43	0.43
Max Drawdown	57.46%	51.47%	49.98%	50.70%	48.01%	52.70%	49.51%

Source: Performance Comparison using MSCI World Index as a benchmark. Reprinted from The Dimensions of Quality Investing: High Profitability and Low Investment Smart Factor Indices, by Scientific Beta An EDHEC-Risk Institute Venture, Retrieved from http://www.scientificbeta.com/#/. Copyright 2015 by ERI scientific Beta

on two factors that include high profitability and low investment. For the index to be implemented using this method it requires 50% of companies achieving high profitability and 50% achieving low invvestmennt. However a limitation exists because from the companies with KAQA only three are listed in Tadawul but luckily during the next award cycle more companies will win the award and so TA`SI-10 Quality Index can be implemented.

REFERENCES

- A. N., F. G., K. G., and N. G. (2015, March). The Dimensions of Quality Investing: High Profitability and Low Investment Smart Factor Indices. Retrieved from ERI Scientific Beta : http://www.scientificbeta.com/#/.
- Advanced Electronics Company (n.d.). About AEC: Quality at AEC. Retrieved from Advanced Electronics Company Web site: http://www.aecl.com/About-Us/Quality.aspx
- Bloomberg (n.d.). Tasawul All-Share TASI Index. Retrieved from Bloomberg L.P: www.bloomberg.com.
- Currier andChet. (2006, March 23). Quality' the new word in investing: [National Edition]. Retrieved from ProQuest: http://search.proquest.com/docview/330426871/abstract?accountid=32779.
- Garvin, D. A. (1991, November). How the Baldrige Award Really Works. Retrieved from Harvard Business Review: https://hbr.org/1991/11/how-the-baldrige-award-really-works.
- King Abdulaziz Quality Award (n.d.). Retrieved from Saudi Standards, Metrology and Quality Org.: http://www.saso.gov.sa/en/about/Pages/kaqa.aspx.
- News, B. B. (1995, February 21). Baldrige quality award winners offer high returns // INVESTING: Over five years, $12,000 put in these companies would yield $23,016.: [MORNING Edition]. Retrieved from ProQuest: http://search.proquest.com/docview/272751991/fulltext?accountid=32779.
- S. K. (2012, April 14). Quality investing for wealth creation. Retrieved from The Hindu BusinessLine: http://www.thehindubusinessline.com/portfolio/quality-investing-for-wealth-creation/article3314975.ece.
- Saudi Press Agency (n.d.). Winning companies to receive King Abdulaziz Quality Award on April 13. Retrieved from Spa: http://www.spa.gov.sa/print.php?id=651246.
- Tadawul (2014, November 20). Market News. Retrieved from http://www.tadawul.com.sa/wps/portal/!ut/p/c0/04_SB8K8xLLM9MSSzPy8xBz9CP0os3g_A-ewlE8TlwN_D38LA09vV7NQP8cQQ3dnA_3g1Dz9gmxHRQBe6u5m/?x=1&PRESS_REL_NO=3912.
- Tadawul (n.d.). About us. Retrieved from Tadawul: http://www.tadawul.com.sa/wps/portal/!ut/p/c1/04_SB8K8xLLM9MSSzPy8xBz9CP0os3g_A-ewlE8TlwODYFMDA08Tn7AQZx93YwN3I_3g1Dz9gmxHRQB7GACB/.

Financing Sources for Subnational Governments in the Fiscal Decentralization of the U.S. and China: Convergence and Difference

Ping Zhang
Ph.D, Associate Professor, Vice Director,Tianjin University of Finance and Economics, China

1. INTRODUCTION

Fiscal Decentralization involves the delegation to multiple levels of government the responsibilities to perform the functions of public sectors. Most countries of the world have at least some degree of fiscal decentralization. The United States and China follow the same general principles in the practice of fiscal decentralization. However, due to the different backgrounds of law and history, the precise structure of fiscal decentralization of the U.S. and China differs.

This paper compares the financing sources for subnational governments in the fiscal decentralization of the U.S. and China. It proceeds as follows. After the introduction, the first section outlines similarities in the distribution of revenues by levels of government. Due to differences of laws, history and backgrounds, the revenue from public bonds has become the principal difference in the financing sources for subnational governments of the U.S. and China. The second section discusses this difference in specifical details. The third section presents some suggestions based on comparative research, to secure the financing sources of and concurrently minimize the fiscal risks for Chinese subnational governments. The final section summarizes the main findings.

2. CONVERGENCE OF FINANCING SOURCES FOR SUBNATIONAL GOVERNMENTS OF THE U.S. AND CHINA

2.1 The Structure of Government Revenue across the Countries

The U.S. government, in an effort to reduce public confusion over the large variety of budgetary concepts, has adopted the concept of a so-called 'unified budget'. That means the on-budget and off-budget revenue are consolidated, which results in a single measure of the fiscal status of the government, based on the sum of all government activity. China, since the 1990's, has established a system of multiple budgets comprising: the general public budget, the government fund budget, the state-owned capital operating budget and the social security fund budget. The general public budget involves all the taxes and non-tax revenue.

Generally, tax revenue and non-tax revenue are the basic sources of public finance revenue for all levels of government in both the U.S. and China (Table 1& Table 2). Tax revenue as a share of government revenue is 83.9% (computed from data of Table 1) in the United States, and it is 85.0% (computed from data of Table 2) in China.

2.1.1 tax revenue

Since different states in the U.S. may have diverse taxes approved by their voters, it is hard to sum all taxes by sources among the U.S. subnational governments. For example, personal income taxes, sales and use taxes, and corporate income taxes are the three main sources of tax revenue (Figure 1) for the state of California.

In China the provinces do not control the actual legislative taxing power. Hence, legislations for taxation in China are rare. Subnational governments have no power to levy taxes except the administration authority. The sources of tax revenue for China's subnational governments have been shown in Table 2.

Table 1. United States Federal, State and Local Government Revenue

				Fiscal Year 2014 in $ billion
	Federal	State	Local	Total
Income Taxes	1,715.3	354.9	37.2	2,107.4
Social Insurance Taxes	1,023.5	450.6	50.4	1,524.5
Ad Valorem Taxes	183.5	508.4	564.5	1,256.4
Fees and Charges	0.0	190.9	265.2	456.1
Business and Other Revenue	99.2	153.7	225.6	478.5
Total	3021.5	1658.5	1142.9	5822.9

Source: http://www.usgovernmentrevenue.com/yearrev2014_0.html

Table 2. Main Items of Public Government Revenue for China's Central and Subnational Governments (2014)

			($ billion)
Item	Total Government Revenue	Central Level	Subnational Level
National Government Revenue	14037.0	6449.3	7587.7
Total Tax Revenue	11917.5	6003.5	5914.0
Domestic Value Added Tax	3085.5	2110.3	975.2
Domestic Consumption Tax	890.7	890.7	
VAT and Consumption Tax from Imports	1442.5	1442.5	
VAT and Consumption Tax Rebate for Exports	-1135.6	-1135.6	
Business Tax	1778.2	6.9	1771.3
Corporate Income Tax	2464.2	1581.3	882.9
Individual Income Tax	737.7	442.6	195.1
Resource Tax	108.4	4.5	103.9
City Maintenance and Construction Tax	364.5	18.3	346.2
House Property Tax	185.2		185.2
Stamp Tax	154.0	64.7	89.3
Stamp Tax on Security Exchange	66.7	64.7	2.0
Urban Land Use Tax	199.3		199.3
Land Appreciation Tax	391.5		391.5
Tax on Vehicles and Boat Operation	54.1		54.1
Tax on Ship Tonnage	4.5	4.5	
Vehicle Purchase Tax	288.5	288.5	
Tariffs	284.3	284.3	
Farm Land Occupation Tax	205.9		205.9
Deed Tax	400.0		400.0
Tobacco Leaf Tax	14.1		14.1
Other Tax Revenue	0.0		0.0
Total Non-tax Revenue	2119.5	445.8	1673.7
Special Program Receipts	371.1	40.6	330.5
Charge of Administrative and Institutional Units	520.6	36.6	484.0
Penalty Receipts	172.2	8.9	163.3
Operating Income from Government Capital	317.6	203.0	114.6
Income from Use of State-owned Resources(Assets)	436.7	17.9	418.8
Other Non-tax Receipts	301.2	138.8	162.4

Source: China statistical yearbook 2015

2.1.2 non-tax revenue

Non-tax revenue of the U.S. subnational governments includes fees and charges, business taxes and/or fees and other revenue (Table 1). Different states might have different statistical definitions for other revenue. Non-tax revenue of China's subnational governments includes special program receipts, charge of administrative and institutional units, penalty receipts, other non-tax receipts (Table 2).

2.1.3 sources of Chinese government funds

Since China's governments at all levels administer multiple budgets, there is a large amount of fund revenue (usually also called as non-tax revenue in reality) except the general public revenue which only tax and non-tax revenue are included. The largest fund revenue for central government is from the account of special national debt management fund. And the most important item of fund revenue is the transfer of use rights of state land which is 81.5%

Figure 2. Shares of Major Taxes for Each Level of China's Subnational Governments (%)

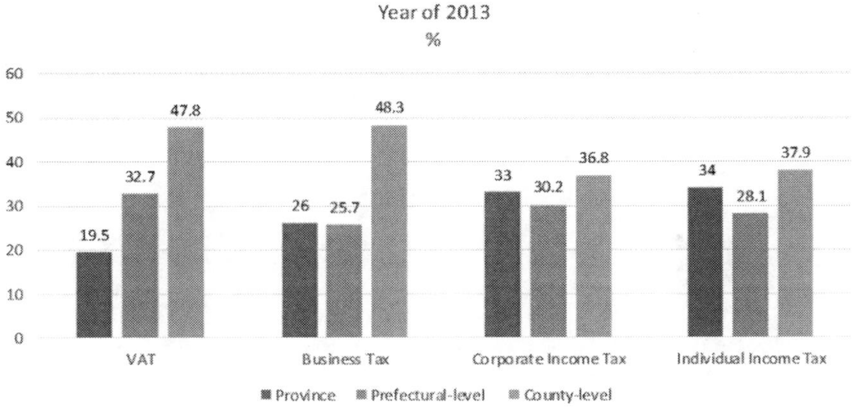

Source: Fiscal yearbook of 2014, Ministry of Finance of the People's Republic of China

of total fund revenue at subnational levels and more than one half of general public revenue of subnational governments[1].

Figure 1. Sources of General Fund Revenue of California

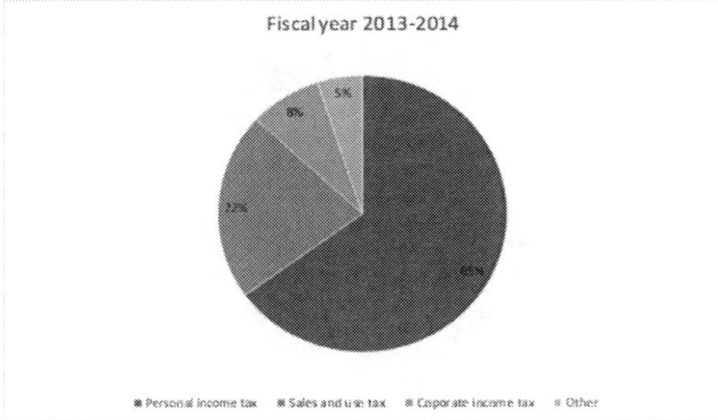

Source: California state controller's office

2.2 Assigning Tax Sources to Different Levels of Government

2.2.1 overview

The United States federal government relies most on personal income taxes. Both the federal and state governments impose personal income tax and corporate income tax. In California, in addition to personal and corporate income taxes, sales and use taxes are also important sources of revenue. These three items constitute 95% of general fund revenue of California. Today, California's counties, cities, schools and special districts depend on the property tax as a primary source of revenue. These funds were allocated as follows: counties 15 percent, cities 12 percent, schools (school districts and community colleges) 54 percent, and special districts 19 percent.[2]

China's subnational governments have hardly any discretionary power to modify taxation. The 1994 tax reform and some complementary institutions thereafter assign consumption tax, VAT and consumption tax from imports, tariffs, and vehicle purchase tax to the central government (see Table 2). House property tax, urban land use tax, land appreciation tax, tax on vehicles and boat operation, farm land occupation tax, deed tax and tobacco leaf tax are assigned to subnational governments (see Table 2). Most of these subnational taxes are related to the use of land and amount to only 18.7% of the total subnational taxes. The remaining 81.3% of subnational taxes have to be shared with the central government. The shared percentage of main taxes for each level of subnational governments is shown in Figure 2. And it is easy to find out both subnational governments of the U.S. and China bear fiscal imbalance pressure, as to the specific circumstances, the pressure of China is more serious (Figure 3 and Figure 4).

2.2.2 general principles of tax sharing system

Although China's subnational governments lack dedicated tax resources and face more testing fiscal pressures than their counterparts in the US, the general principles for assigning tax resources to different levels of government are similar in both countries. Generally speaking, households and corporations can move more easily from one locality to another than they can from one state or province to another or than they can from one country to another. For this reason, central governments are best placed to collect taxes on the more mobile sources of tax revenue. State or

1 China Statistical yearbook 2014, Ministry of finance of the People's Republic of China, http://yss.mof.gov.cn/2013qgczjs/201407/t20140711_1111744.html.

[2] California Property Tax An Overview, Publication 29, July 2015

Figure 3. China's Government Finance: Central vs. Subnational (2014) Billions of Yuan

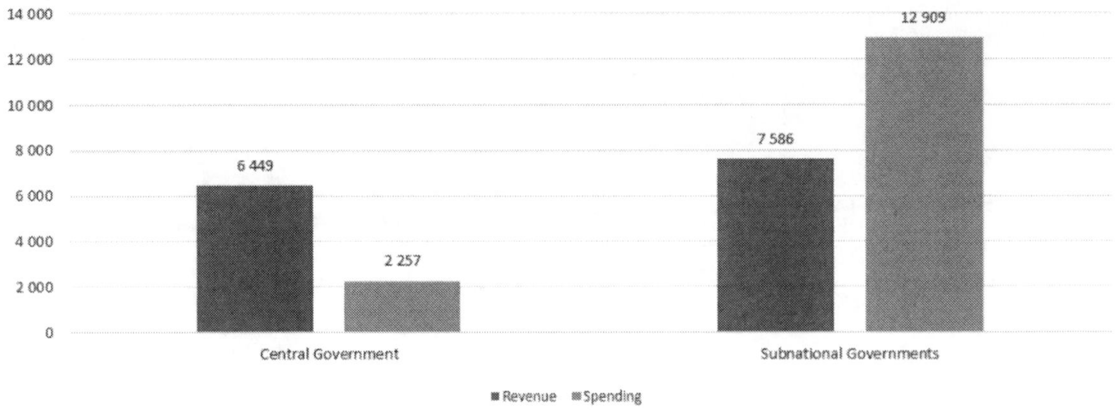

Source: Ministry of finance of the People's Republic of China, http://www.mof.gov.cn/zhengwuxinxi/caizhengxinwen/201503/t20150317_1203481.html

Figure 4. U.S. Government Finance: Federal vs. Subnational (2015) Billions of Dollars

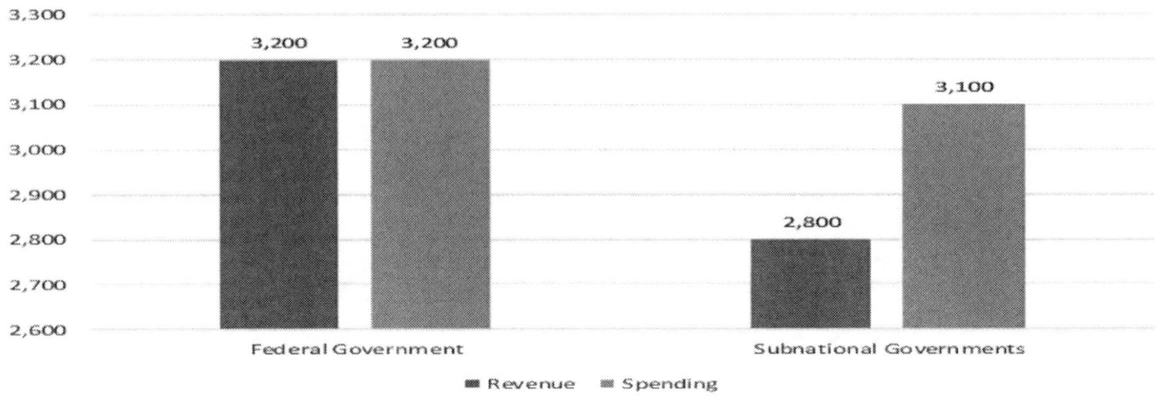

Source: http://www.usgovernmentrevenue.com/, http://www.usgovernmentspending.com/total

provincial governments levy taxes on less mobile tax sources; localities focus on the least mobile sources of revenue.

2.3 Intergovernmental Transfers

Because lower levels of governments have narrow tax bases, a decentralized fiscal system usually incorporates revenue transferred from higher to lower levels of government. These transfers include general purpose grants, which allow the recipient government to use the funds unrestrictedly; and special purpose grants, which allow the recipient governments to use the funds freely within a special program or purpose area. Sometimes, special purpose grants could require recipient governments (and agencies) to contribute to the cost of specific programs as a condition for receiving transfers from higher levels of government. In recent decades, US federal aid to state and local governments has soared, thereby increasing their reliance on federal aid for the financing of certain subnational government functions. Federal grants-in-aid have become an important financing source for subnational governments, for example, contributing nearly half of general revenue in Mississippi[3]. Most Federal grants-in-aid are program grants (so called special purpose grants).

In China, the amount of general purpose grants has exceeded special purpose grants since 2010 (Figure 5). That means subnational governments increasingly enjoy more flexibility and security in the revenue receive from the central government. China's subnational governments rely on intergovernmental transfers from central government more heavily than the U.S. (Figure 6).

[3] *B William and J Luppino-Esposito, 'Increased federal aid to states is a long term trend', State Budget Solutions, March 26, 2014.*

Figure5. China's Central Grants to Subnational Governments (Billions of Yuan)

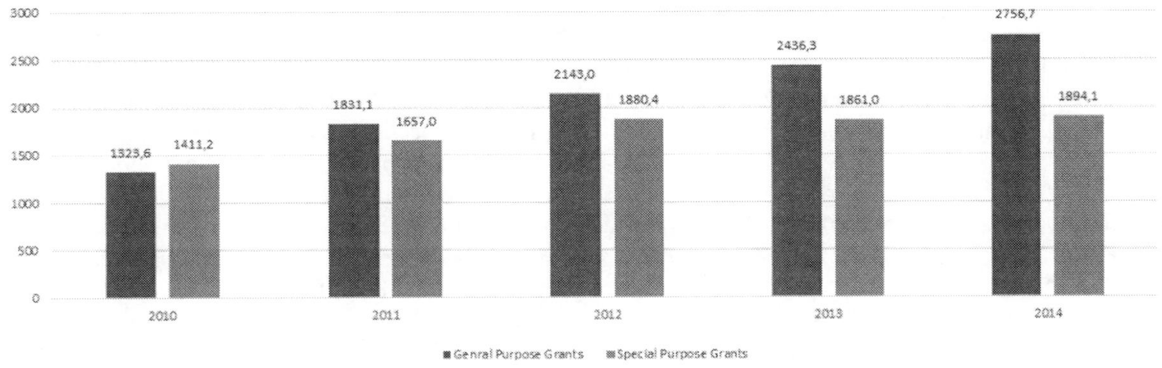

Source: Ministry of finance of the People's Republic of China

Figure 6. The Share of Chinese Intergovernmental Transfers in Subnational Government Revenue (%)

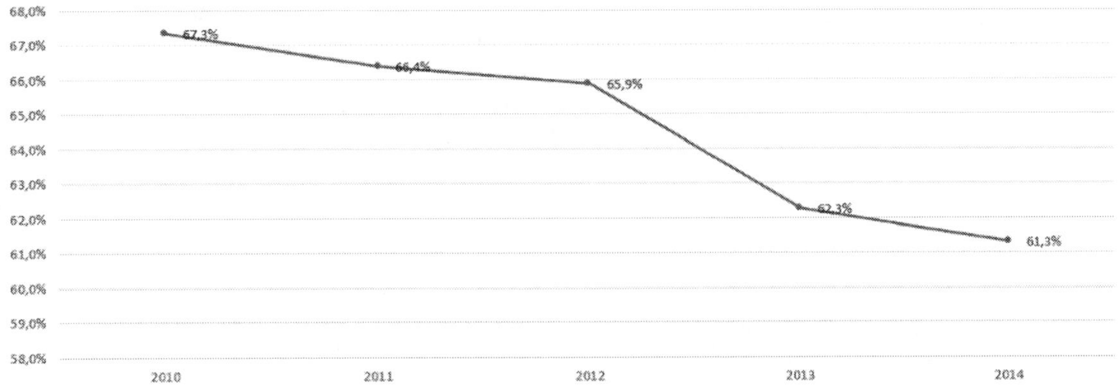

Source: Ministry of finance of the People's Republic of China

2.4 Public Bonds

Tax revenue and intergovernmental transfers can resolve much of the funding demand for current expenditures for both subnational governments of the U.S. and China. When it comes to the capital expenditures of subnational governments, however, many U.S. states empower local governments to issue bonds to finance their capital spending. Municipal bonds usually are issued for the purpose of financing specific infrastructure needs of the issuing municipality. The U.S. municipal bond market amounted to \$3.63 trillion in the third quarter of 2014[4], unique in the world for its size, liquidity, legal and tax structure, as well as the bankruptcy protection afforded by the U.S. Constitution.

China's subnational governments had not been proscribed by the law from issuing bonds. China's subnational governments have to borrow from the public in other indirect ways. A revised budget law has taken effect since the beginning of 2015, under which all provinces, prefectural-level units and municipalities have the right to issue public bonds within the restricted amounts approved by the State Council. Even after these desirable changes, the subnational governments still contend with considerable pressure in funding their capital spending, given the rapid urban development of China. So unlike the United States, subnational public debt 'not federal debt is the main pressure point for Chinese government debt.

2.5 Summary

Generally, administrative capability and mobility are the two basic factors that determine the assignment of taxes in both the U.S. and China. Both central governments support the activities of subnational governments through a variety of grants and/or transfers. Tax revenue and intergovernmental transfers

[4] *Federal Reserve data released on Dec 11, 2014.*

predominantly finance current public spending. Both U.S. and China's subnational governments have to issue bonds for their capital spending. Due to the distinctions of laws, history and backgrounds, the revenue from public bonds becomes the prominent difference of financing source for subnational governments of the U.S. and China.

3. DIFFERENCE OF FINANCING SOURCES FOR SUBNATIONAL GOVERNMENTS OF THE U.S. AND CHINA

3.1 Characteristics of China's Subnational Governments Bonds

3.1.1 off-budget financing sources dominate

Subnational governments employed local government financing vehicles (LGFVs) and other state-owned enterprises (SOEs) effectively to borrow indirectly in off-budget ways, for example, from policy banks, commercial banks and, more recently, trust companies and the corporate bond market. Therefore, LGFVS and SOEs hold the largest share of outstanding government debt. According to investigations by the National Audit Office of the PRC, bank loans accounted for around 80% of total subnational government debt by 2010. With the awareness of the repayments risks of LGFVs and the huge burden of accumulating local debt, the central government introduced some policies, in effect, to control the commercial bank loans borrowed by LGFVs. That compelled subnational governments to look elsewhere for capital. Commercial bank loans decreased to 56.56% of total subnational government debt by the end of June, 2013. By contrast, many other financing sources like BT (Build-Transfer) and trust financing, which were never used by subnational governments, emerged. Except the traditional financing source -- bank loans, BT and trust financing are the top two items that contributed to the subnational debt.

3.1.2 Fiscal Risks and Financial Risks Are Highly Related

LGFVs are largely set up, owned, and operated by the local governments. Although LGFVs may also generate revenue and some may be operated on a commercial basis, they engage in economic activities that are fiscal in nature, borrow from financial market. And the governments explicitly or implicitly share the LGFVS' debt servicing responsibilities, and sometimes subsidize their losses. Because of the existence of LGFVs, financial risks and fiscal risks are linked.

Subnational Debts Exposed to Shadow Banking Activity. China's banks and trust companies have developed a range of wealth management products through which they provide loans to subnational governments in recent years. They value 'off balance sheet lending', as it allows them to avoid restrictions on loan-to-deposit ratios. It also allows them to speculate on commodity prices, invest in real estate projects or provide loans at high interest rates to enterprises and local government projects that face borrowing restrictions in the formal sector. Due to the typically less-transparent nature of these products, the pricing of these loans might not reflect the true underlying risk.

Mismatched Risk of Wealth Management Products. The worst risk that comes to mind involves the wealth management products, especially infrastructure trust products, that borrow short (typically less than six months) and lend long (by investing in bonds and loan products longer than three years). The mature infrastructure trust products amounted to an estimated ¥5.3 trillion in 2014[5]. This portion of the wealth management products may face rollover risk if investors are unwilling to continue to finance the underlying projects by buying the new products offered upon the maturity of the old.

[5] *Wind data, http://www.wind.com.cn.*

Table 3. Subnational Government Debt in the United States (Amounts in $ billion; FY2015) Top 5

FY2015	State &Local Debt	Gross State Product	State &Local Debt against Gross State Product %
New York	359.9	1,389.7	25.9
Rhode Island	12.7	56.8	22.4
Kentucky	42.8	195.7	21.9
Pennsylvania	142.1	684.1	20.8
South Carolina	40.1	196.0	20.5

Source: http://www.usgovernmentspending.com/state_spend_gdp_population

High Borrowing Cost. LGFVs borrowing costs are generally around double that of the central government. Government bond yields average around 3 percent in nominal terms, while bank lending rates to LGFVs have tended to exceed the benchmark lending rate (by around 1–2 percentage points). Typical LGFVs debt, therefore, would carry an interest rate of 6-8 percent, or more than double that of the central government. Trust loans tend to be more expensive. The rates can range from as low as 6½ percent to above 10 percent. Overall, the shift to bank market and shadow bank borrowing by LGFVs almost surely has raised their interest costs.

3.2 The U.S. State Bonds

As analyzed above, unlike China, in the United States federal government debt, not subnational government debt, raises most concern. Table 3 shows that New York has the highest ratio of state and local debt to gross state product.

3.2.1 New York state and New York city can serve as examples

The two basic types of New York State Bonds are general obligation bonds and revenue bonds. In New York State, general obligation bonds are voter-approved. Principal and interest are secured by the full faith and credit of the State of New York and are supported by the State's unlimited taxing power. Revenue bonds are guaranteed by the specific revenue generated by the issuer; issuance does not require voter approval. The most common issuers of revenue bonds are transportation systems, hospitals, power systems, sewer systems and water systems. These issuers could generate revenue from providing services to the public for the purpose of paying the bonds.

3.2.2 the operating mechanism of New York city municipal water bonds

Nine participants are involved in the operating pattern of New York City municipal water bonds. They are the water board, the depository trust company, the New York City municipal water finance authority, the bond counsel, independent advisors, the co-senior manager, the book-running senior manager, commercial banks and the public. The water board, comprising of seven people, authorizes the New York City municipal finance authority to issue municipal water bonds. The authority is responsible for announcing the details of the bonds, selling the bonds to the public and commercial banks. The bond counsel will advise the authority during the whole issuing process. The co-senior manager and book-running senior manager are serving for the specific transactions. Independent advisors could recommend the yield rate for short-maturity bonds. Yields on bonds of longer maturity must be negotiated between all the involvers. In order to service the bonds, the water board could collect fees from the public for its use of water and sewer services. And the board also needs to deposit an agreed sum, at least equal to the minimum monthly balance sufficient to pay the principal and interest due on the bonds, with the depository trust company. That deposit ensures the creditworthiness of the issuer to the investor. To guarantee sufficient cash flows to avoid default, the New York City municipal water finance authority invests its revenues in eleven ways. And there are strict quantitative standards for each avenue of investment.

4. SUFFESTIONS OF COMPARATIVE RESEARCH

4.1 To Secure Revenue Sources for Subnational Governments

4.1.1 introducing taxes on residential property

The introduction of taxation on residential property has been under consideration since 2003 as a means of improving subnational tax revenue. China has both property and land taxes but they are not applied to residential real estate. The government has considered introducing a new residential property tax that might be used either to raise local government revenue or to reduce the reliance of local governments on transfers from higher levels of government. Although the tax on property transactions is one of the few taxes for which a provincial government can set the rate in China, the current property tax is levied only on commercial property. By contrast, the U.S. local governments rely heavily on the property tax, which applies mainly to residential and commercial properties. So if China introduces a nationwide residential property tax the pressure on local government fiscal revenue could be relieved. And the tax rate need not be uniform across the country, but could vary in consideration of the differences among prefectural-level cities and their districts.

4.1.2 assigning additional taxing authority to subnational governments

The U.S. fiscal federalism is based on local autonomy. Local governments have a high degree of autonomy and independence in their recourse to taxation. But China differs from the U.S. because of the peculiarities of history, economic and social development and the political system. China operates under a centralized unitary system; the central government exerts absolute political dominance, and the local governments have limited autonomy, especially when it comes to taxation. According to the US practice, the basic taxation law, which deals with the tax sharing system between different levels of government, should equip local governments with more power to tax, thereby contributing to stable fiscal decentralization.

4.2 To Minimize the Fiscal Risks of Subnational governments

4.2.1 delinking fiscal risks from financial risks

On one hand, the governments should relieve the risks from the dimension of space, which implies governments should find the appropriate financing sources of bond revenue. Therefore, the Central government should introduce measures to prevent subnational governments from using shadow banking activity inappropriately to raise funds for projects. To bar local governments from using trust loans to raise funds for projects those are not fully covered by cash flows. China's Subnational governments need to learn from the U.S. experience with municipal bonds.

On the other hand, the governments should introduce measures that relieve the risks deriving from time, which implies governments should structure the maturity of bonds appropriately. Since 2014, China's central government allows LGFVs to issue bonds to rollover outstanding credits from banks so as to avoid defaults. During 2015, the Central government has arranged for the refinancing of 2 trillion of maturing debt by allowing provincial governments to sell bonds for the purpose of reducing the cost, and extending the maturity, of debt by replacing short-term, high interest rate bank loans with cheaper, longer-term municipal bonds.

4.2.2 changing the growth model of China's subnational economies

China's subnational governments must transform their economies to depend less on the current investment-led growth model to one based more on balanced domestic demand with an emphasis on services. The economy growth should be assessed by various indicators. Not only the local GDP growth rate, but also the civil private economy active degrees, the safety of local public finance should be taken into account.

4.2.3 reducing problems of subnational governments associated with moral hazard

As long as subnational governments believe, even implicitly, that their debts are guaranteed by the central government and if they put themselves in desperate circumstances the central government will have to come to their aid, those governments will lack the necessary incentives to manage their finances responsibly. Furthermore, if bankers also believe the Central government stands ready to ride to the rescue in the face of every misfortune, funds from the formal banking system will continue to flow freely into risky sectors favored by local officials. It may be premature and even inappropriate to talk about the possibility of a bankruptcy law for subnational governments. But by sending a clear message that the Central government will no longer provide bailouts for unregulated and uninsured investments, and those investors will experience capital losses for poor and irresponsible decisions; regulatory agencies could go a long way towards curbing risky activities by encouraging both the official and shadow banking sectors to price risks more realistically.

4.2.4 imposing hard budget constraints on subnational governments

All levels of government should face hard budget constraints on their public activities to rein in uncontrolled spending and limit the current expansion of credit and debt. Put differently, meaningful measures on budget law coupled with the imposition of tough financial discipline on local governments are an urgent task.

5. CONCLUSIONS

The United States and China have similar financing sources for subnational governments and basically embrace similar principles in managing different financing sources under fiscal decentralization. One can observe a trend towards convergence in the fiscal decentralization of the U.S. and China. The dominant difference in the financing sources for subnational governments of the U.S. and China is in the reliance on public bonds. At all levels of China's governments, officials should study the enabling legislation for public bonds and the associated institutional design of the U.S so as to limit the risks involved in the issuance of subnational debt and prevent interlocked fiscal and financial risks from leading to systemic risk.

REFERENCES

- Yuanyan Sophia Zhang and Barnett, S. (2014). Fiscal Vulnerabilities and Risks from Local Government Finance in China. IMF Working Paper.
- Xiao Wang and Herd, R. (2013). The System of Revenue Sharing and Fiscal Transfers in China. OECD Economics Department Working Papers No.1030.
- Chi Lo (2013). What Ponzi Scheme? *The International Economy,* 32-33, 66-67.
- Shalendra D. Sharma (2015). China's Debt Woes: Not Yet a 'Lehman Moment. *Journal of East Asian Economic Integration, Vol. 19, No.1,* 99-114.
- Hehui Jin, Yingyi Qian, and Barry R. Weingast. (2015). Regional Decentralization and Fiscal Incentives: Federalism, Chinese Style. *Journal of Public Economics, 89,* 1719 -1742.
- Shaun Breslin (2014). Financial Transitions in the PRC: Banking on the State? *Third World Quarterly, Vol. 35, No.6,* 996-1013.
- Junxue Jia, Qingwang Guo, and Jing Zhang (2014). Fiscal Decentralization and Local Expenditure Policy in China. *China Economic Review, 28,* 107-122.
- Li Han and James Kai-Sing Kung (2015). Fiscal Incentives and Policy Choices of Local Governments: Evidence from China, Journal of Development Economics, 116, 89-114.
- Guess, G. M. and Jun Ma. (2015). The Risks of Chinese Subnational Debt for Public Financial Management. *Public Administration and Development, 35,* 128-139.
- Yanrui Wu (2014). Local Government Debt and Economic Growth in China. BOFIT Discussion Papers.
- Zhiwu Chen (2015). China's Dangerous Debt: Why the Economy Could Be Headed for Trouble. *Foreign Affairs, 94,* 13-18.
- Ehtisham Ahmad (2014). Public Finance Underpinnings for Infrastructure Financing in Developing Countries. The London School of Economics and Political Sciences.
- Joshua E. Greene (2012). Public Finance: An International Perspective, World Scientific Publishing Co. Pte. Ltd.
- Yuhui Liu (2014). Local Government Debt Management Solutions. *China Economist, Vol.9, No.5,* 90-97.
- Tiankai Wang, Shields, P., and Yangmei Wang (2014). The Effects of Fiscal Transparency on Municipal Bond Issuances. *Municipal Finance Journal, Vol. 35, No.1,* 25-44.
- Yilin Hou (2015). Local Government Budget Stabilization: Explorations and Evidence. Springer International Publishing.

Inter-Regional Economic Relation, Industrial Structure and Pollution Reduction 'Benchmark Synergy'

Wei Zeng
Postgraduate, Tianjin University of Finance and Economics, China

1. INTRODUCTION

By the end of 2015, Beijing issued the red alert of smog twice within every 10 days. In fact, Beijing has adopted a series of measures such as improving the energy structure, relocating polluting enterprises and improving vehicle emissions standards to reduce smog emission. After years of efforts, sulphur dioxide, nitrogen oxides and smoke (powder) dust emissions for each 10,000 yuan GDP of Beijing have been reduced to 14%, 24% and 15% of the national average. However, as the surrounding areas are mostly high emission zone, environmental pollution problems have not yet been fundamentally resolved. Although the administrative means of production restriction and emission reduction has been taken to improve the environment, which is nevertheless not a permanent solution. The key is to find an inter-regional pollution reduction method to solve the problem fundamentally.

There is obvious difference between the Chinese and foreign studies on how to realize the regional pollution reduction coordination. Foreign literatures pay more attention to the role of emission reduction contracts in pollution control, so they prefer using game theory to study the stability and the actual effect of emission reduction contracts. They study the factors affecting the stability of pollution reduction contract such as new entrants behavior and participation strategy. Diamantoudi (2006) focused on the effect of participants' number, and the pollution reduction effect of collaborative was evaluated by Rypdal (2007). Most of the Chinese researches study the coordination of inter-regional governmental relations by qualitative analysis. Li Xuesong (2014) suggested that strengthening the punishment and supervision can enhance regional pollution reduction synergy, Zhang Yanbo (2015) suggested innovating coordination mechanism between local government, and Huangbin Huan (2015) suggested restructuring the system's balance of the state, market and society, Cui Jing (2014) put forward to the public participation.

But so far, the relationship between pollution reduction synergy, regional economic relation, and industrial structure, is still less involved in the existing research. If we set the region with high level of economic development and low emission level of pollutants as a benchmark to follow, the regional pollution reduction 'benchmark synergy' can be regarded as the consistency and synchronization of high pollution area with the 'benchmark' in the direction and range of pollution reduction. The ultimate goal is to make the entire region to achieve a 'win-win' relationship between economy and environment. In theory, the strengthened economic relation as well as the improved correlation of industrial structure indicates that regions are more interconnected. So it is easier to form a community of interests, and is of great help to reduce the pollution externality and to enhance the inter-regional pollution reduction 'benchmark synergy'.

It is particularly important for Beijing-Tianjin-Hebei Region to prove whether there exists a relationship between inter-regional economic relation, industrial structure and pollution reduction benchmark synergy. As shown in Figure 1, in view of economic development, the per capita GDP of Beijing reached 93 thousand yuan in 2013, up to the income level of moderately developed countries. From the perspective of regional industrial structure, the secondary industry of Beijing accounts for only 22%, while the proportion of tertiary industry exceeds 75%. Beijing has entered the post-industrial stage in Chenery's industrialization stage theory. From the emission level, ten thousand GDP of sulphur dioxide, nitrogen oxides and smoke (powder) dust emissions are far below the national average level. Therefore, Beijing overally can be regarded as the 'benchmark' area for its well-coordinated development of economy and environment. Comparatively, the neighboring city, Tianjin, with approximately the per capita GDP of 100 thousand yuan, is still in the high-end equipment manufacturing based economy. And the proportion of Tianjin's secondary industry is about 50%. As to Hebei Province, the per capita GDP is only about 39 thousand yuan, its economic growth heavily dependent on the heavy industrial and is still in the middle stage of industrialization of Chenery's industrialization stage theory. At the same time, Hebei Province is a typical highly-polluted area as the sulphur dioxide, nitrogen oxides and smoke (powder) dust emissions level is 1.4 times, 1.7 times and 2.3 times of the national average level. The huge difference in the level of economic development and industrial structure causes difficulties for the Beijing-Tianjin-Hebei Region's collaboration of pollution reduction. With the implementation of Beijing-Tianjin-Hebei Region Coordinated Development Strategy, the inter-regional economic relation will be more interconnected and strengthened than ever. At the same time, for Hebei Province, industrial structure optimization and upgrading will be further accelerated, and the inter-regional industrial complementarity will be enhanced. So, theoretic and empirical analysis will be explored in this study to examine whether the Beijing-Tianjin-Hebei Region will have a 'benchmark synergy' effect in pollution reduction with the change of inter-regional economic relation and industrial structure. The features of this paper are:

(1) Exploring a new approach to realize the inter-regional cooperative reduction of pollution from the point of view of economic relations and industrial structure.

(2) Proving and testing from both theoretical and empirical ways that the pollution reduction 'benchmark synergy' effect has a monotonic increasing relation with the intensity of economic relation, and a 'inverted U' type relationship with the industrial structure isomorphism.

(3) Providing a new idea to realize the inter-regional collaborative governance of environmental pollution based on the research results.

2. THEORETICAL PROOF

As shown in Figure 2, A, B, C three areas will be considered. The production process of A area has two kinds of production factors x_1 and y_1. Beside, x_1 is clean factor, y_1 is pollution factor, and the two factors can be replaced by each other, the marginal rate of technical substitution is $\frac{\partial y_1}{\partial x_1}=-\frac{\partial F}{\partial x_1}\Big/\frac{\partial F}{\partial y_1}$. We assume that he production function of A area is $F(x_1,y_1)$, so $\frac{\partial F}{\partial x_1}>0$ and $\frac{\partial F}{\partial y_1}>0$, what's more $\frac{\partial^2 F}{\partial x_1 \partial x_1}<0$, $\frac{\partial^2 F}{\partial y_1 \partial y_1}<0$. The negative value of two order partial derivation means the marginal output of factor input is gradually decreasing. So it also illustrates that, under the condition of the same technical level, the marginal output level $\frac{\partial F}{\partial y_1}$ is gradually increasing with the decreasing of input quantity of pollution factor y_1. The profit function for the A area is $\pi_A=p_1F(x_1,y_1)-c_{A1}x_1-c_{A2}y_1$, where p_1 is the unit price of goods produced in the A area, c_{A1} is the unit cost of clean factor, and c_{A2} is the unit cost associated with pollution factor. Beside c_{A2} is composed of two parts. One part is the unit price of the pollution factor, and the other part is the loss of local environment caused by the input of unit pollution factor. It is obvious that the optimal input of pollution factor y_1 for A is $\frac{\partial F}{\partial y_1}=\frac{c_{A2}}{p_1}$. Only when the input of y_1 decreases, can $\frac{\partial F}{\partial y_1}$ has a increase.

Figure 1. Comparison of Economic and Pollutant Emission Indices in Beijing-Tianjin-Hebei Region

(a)

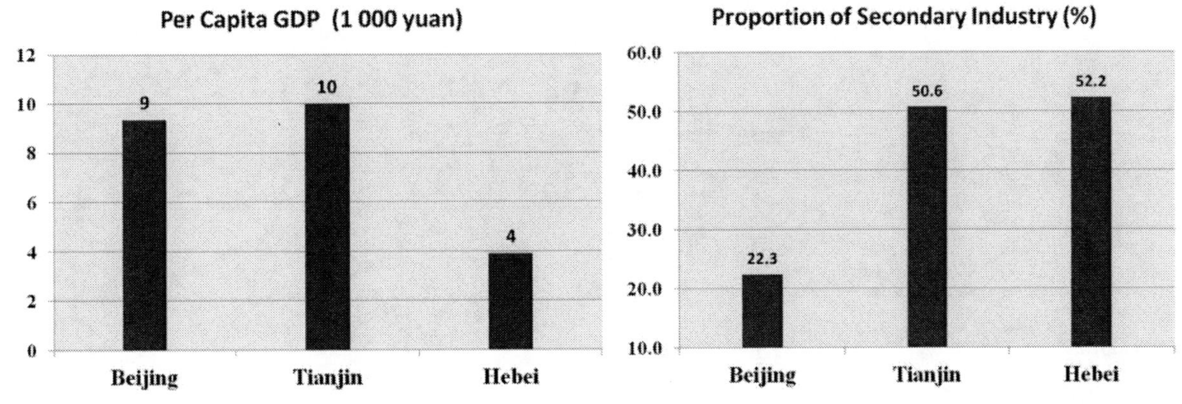

(b)

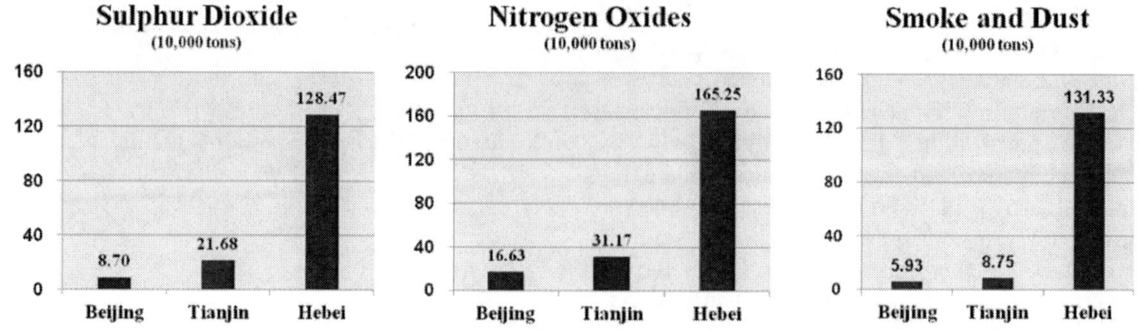

So, if the A area wants to control the input amount of its pollution factor under the condition of constant environmental losses caused by the input of unit pollution factor, there are two paths to choose from:

(1) Reducing the price p_1 of output $F(\bullet)$. In reality, it is often realized by restricting the using number of polluting products (such as 'limited plastic order') and developing clean alternatives or other methods;

(2) Increasing the unit cost c_{A2} of y_1. In reality, it is often realized by the way of taxation or levying sewage charges.

At the same time, B and C two regions with completely alienated industrial structure are assumed. Both two areas are concerned about the economic interests and their own environment quality at the same time, expecting a cleaner mode of production and less emission of pollutants. For the A area, any one of the B and C areas is likely to be regarded as the 'benchmark' region for the well-coordinated development of economy and environment. Assuming the existence of industrial structure alienation between regions, the output of the A area can only be sold to the B area, or can only be sold to the C area, and it mainly depends on which area has the industrial relevance with A.

For the relationship between industrial structure and industry correlation, Yu Dianfans, Gan Chunhui, and Zheng Ruogu (2011) thought it is necessary to reveal the relationship of industrial structure between the two regions firstly. Generally speaking, if the degree of the industrial structure isomorphism of the two regions is too high, it is difficult to form a complementary between industries, and is not conducive to the upgrading of inter-industry correlation among regions. But on the other hand, if the difference of industrial structure between the two regions is too large, it is difficult to form effective industry convergence and it will also adversely affect the degree of inter-regional industry correlation. Figure 3 describes the relationship between the industrial structure and industry correlation, with the increase of the degree of industrial structure isomorphism in the region, the inter-regional industry correlation should be 'first rise and then drop', showing a distribution of 'inverted U' type. In order to simplify the research, assuming that the B area's industrial structure isomorphism with the A area is located in the vertex b of 'inverted U' type curve, at this time the two regions' industry

Figure 2. Regional Relations in the Theoretical Proof

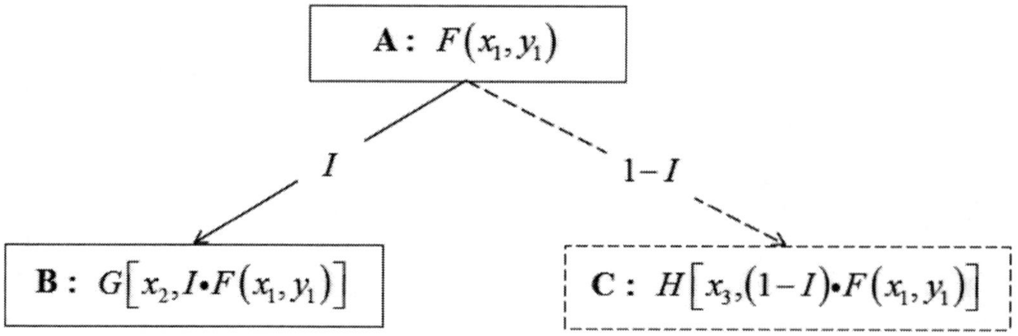

Figure 3. The Relationship between the Industrial Structure Isomorphism and Industry Correlation

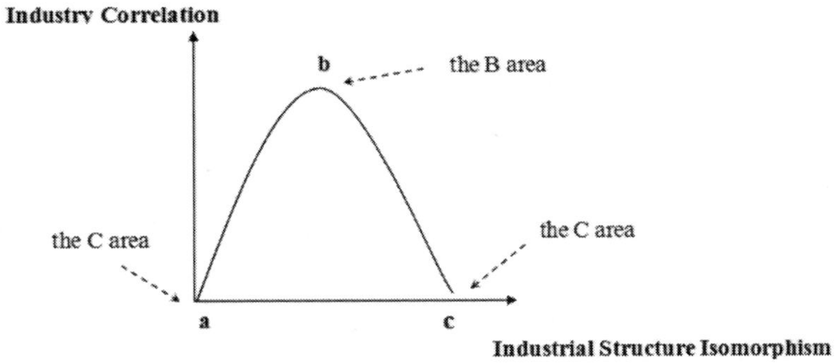

correlation also reached the maximum value. At the same time, it is assumed that the C area's industrial structure isomorphism with the A area is located in a or c, which reflects a lack of industry correlation because of the completely identical or completely irrelevant industrial structure.

As shown in the formula (1) there exists a parameter I for the B and C two areas which obeys the 0-1 Distribution and determines the industry correlation with the A area. Assumed that there are $I_B = 1$ and $I_C = 0$, the production function of the B ares is $G[x_2, \theta F(\bullet)] = G[x_2, F(\bullet)]$ and the production function of the C area is $H[x_3, (1-\theta)F(\bullet)] = H(x_3)$

Among them, x_2 and x_3 can be regarded as the local clean factor input of the B and C two regions.

Analysis of the case that $I_B = 1$. The production function in the B area is $G[x_2, F(\bullet)]$, in which the clean factor x_2 derived from local and the factor $F(\bullet)$ purchased from the A area can be replaced by each other. The marginal rate of technical substitution is $\frac{\partial F(\bullet)}{\partial x_2} = -\frac{\partial G}{\partial x_2} \Big/ \frac{\partial G}{\partial F(\bullet)}$.

For the production function $G[x_2, F(\bullet)]$ of the B area, there also exists $\frac{\partial G}{\partial x_2} > 0$ and $\frac{\partial G}{\partial F(\bullet)} > 0$, as well as $\frac{\partial^2 G}{\partial x_2 \partial x_2} < 0$ and $\frac{\partial^2 G}{\partial F(\bullet) \partial F(\bullet)} < 0$. The expression $\frac{\partial G}{\partial F(\bullet)}$ reflects the strength of economic relation between the A and B two areas. With the increase of the intensity of economic relation, B areas will have more production factors to purchase from the A area, specifically reflecting the gradual increase of $F(\bullet)$. As the $F(\bullet)$ increases, $\frac{\partial G}{\partial F(\bullet)}$ should be gradually declining under the effect of the law of diminishing marginal output. At this time, the profit function of B area is $\pi_B = p_2 G(x_2, F(\bullet)) - c_{B1} x_2 - p_1 F(\bullet) - c_{B2} y_1$, c_{B1} is the costs for the purchase of clean factor from local and c_{B2} is the external impact of the pollution factor input of A to the B area. In order to analyze the B area's expected optimal pollution factor input of the A area, the first-order condition about y_1 for the B region is solved and the result is: $\frac{\partial \pi_B}{\partial y_1} = p_2 \frac{\partial G}{\partial F} \frac{\partial F}{\partial y_1} - p_1 \frac{\partial F}{\partial y_1} - c_{B2} = 0$ then we can obtain

$$\frac{\partial F}{\partial y_1} = \frac{c_{B2}}{p_2 \frac{\partial G}{\partial F} - p_1} \quad (2)$$

As previously analyzed, the pollution factor y_1 of A area is expected to decrease, so $\frac{\partial F}{\partial y_1}$ should be incremental under the condition of the same technical level. From this we can know that the in the A area could be reduced in cases that: (1) The enhanced pollution external impact from the production process of the A area, will make c_{B2} gradually become larger and $\frac{\partial F}{\partial y_1}$ gradually increase, which means that the B area would expect the A area to reduce the input of y_1 otherwise the B area would like to replace the $F(\bullet)$ by using the local cleaning factor x_2; (2) In the case of the constant production factor price c_{B1}, the gradual increase in the purchase price p_1 of the $F(\bullet)$ from the A area or the decrease in the B region's output price p_2 of $G(\bullet)$ will lead to a decrease in the purchase of production from A, which push the A area to decrease the input of y_1. (3) The increase of economic relation of the A and B two area means that there will be more inputs purchased from the A area to the B area. Thus, with the $F(\bullet)$ increasing gradually $\frac{\partial G}{\partial F(\bullet)}$ and decreasing, $\frac{\partial F}{\partial y_1}$ becomes larger, which objectively need the A area to reduce y_1.

At the same time, according to the industry completely opposite hypothesis, when $I_B = 1$, there must be $I_C = 0$. Therefore, the production function of the C area is $H(x_3)$. The production process of the C area has no relationship with the A area, so the C area has no influence on the choice of the A area about the use of clean factor x_1 and pollution factor y_1.

Based on the above analysis, the industry correlation and the strength of economic relation are also the key factors affecting the pollution emission, except for traditional measures such as the restrictions on the use of contaminated products, the development and promotion of clean alternative technologies and products and the collection of sewage charges or sewage tax to enhance the use cost of pollution factor. When a close economic relation and a moderate level of industrial structure isomorphism with the 'benchmark' region is maintained, a cleaner production method can be used to reduce the pollutant emission. In this way, the inter-regional pollution reduction 'benchmark synergy' functions quiet well.

3. EMPIRICAL TEST

3.1 Index Construction and Data Description

Although the reasons for smog are still controversial, existing studies generally believe that it is related to the burning of fossil fuels and emissions from automobile exhaust. In this paper, data of smoke (powder) dust emissions of Beijing, Tianjin and other 11 prefecture-level cities in Hebei Province from 2004 to 2013 is chosen as the proxy variable to reflect the emission of pollutants. Taking into account the level of economic development and pollutants emission in reality, we set Beijing as the 'benchmark' area, and calculate the remaining 12 cities' 'benchmark synergy' of smoke (powder) dust emissions with Beijing. The general correlation coefficient can only calculate the cross section data; therefore, this paper uses the method mentioned by Cerqueira (2009) of constructing a 'benchmark synergy' index $Corr_{bi,t}$ of pollution reduction activities in each period:

$$Corr_{bi,t} = 1 - \frac{1}{2}\left(\frac{(p_{b,t} - \bar{p}_b)}{\sqrt{\frac{1}{T}\sum_{1}^{T}(p_{b,t} - \bar{p}_b)^2}} - \frac{(p_{i,t} - \bar{p}_i)}{\sqrt{\frac{1}{T}\sum_{1}^{T}(p_{i,t} - \bar{p}_i)^2}} \right)^2 \quad (3)$$

In equation (3), $p_{b,t}$ is the smoke (powder) dust emissions of Beijing in the t year, and $\bar{p}_b$ is the average smoke (powder) dust emissions of Beijing from 2004 to 2013; $p_{i,t}$ represents the smoke (powder) emissions of each 12 remaining cities of Beijing-Tianjin-Hebei Region in the t year (i = 1. . .12) , and $\bar{p}_i$ denotes the the average smoke (powder) dust emissions of 12 cities from 2004 to 2013.

Due to the lack of specific trade volume data, the existing Chinese literature mostly use gravity model to reflect the inter-regional economic relation. The gravity model takes into account the factors such as population, economic gross and spatial distance, and the bigger its result is, the stronger the economic relationship is. The specific expression $R_{bi,t}$ of the economic relationship between Beijing and the remaining 12 cities in each period is

$$R_{bi,t} = \frac{\sqrt{S_{b,t}G_{b,t}} \times \sqrt{S_{i,t}G_{i,t}}}{D_{bi}^{2}} \quad (4)$$

In equation (4), $S_{b,t}$ is the population of Beijing in the t year, and $G_{b,t}$ is the gross domestic product (GDP) of Beijing deflated by consumer price index (based period 2004). $S_{i,t}$ is the population of each

12 remaining cities in the t year, and $G_{i,t}$ is the GDP of these cities treated with the same method of $G_{b,t}$. D_{bi} is the nearest highway distance between the city i and Beijing. Finally, based on the time dimension, the normalized method is used to deal with the result of the gravity model, so as to accurately reflect the changes in the economic relationship between the city i and Beijing.

Just as the analysis of the theoretical model, the level of industrial structure isomorphism is closely related to the degree of inter-regional industry correlation. And the industrial structure isomorphism index can make up the lack of input and output data of Beijing-Tianjin-Hebei Region in the process of calculating the degree of industry correlation. The index of industrial structure isomorphism is $H \in [0,1]$ and $H = 1$ means the completely identical industrial structure of the two areas, while $H = 0$ means the completely irrelevant industrial structure. The concrete expression of the industrial structure isomorphism index $H_{bi,t}$ of remaining 12 cities with Beijing in each year is

$$H_{bi,t} = \frac{\sum_{k=1}^{n} X_{bk,t} X_{ik,t}}{\sqrt{\sum_{k=1}^{n} X_{bk,t}{}^{2} X_{ik,t}{}^{2}}} \quad (5)$$

In equation (5), $X_{bk,t}$ is the proportion of Beijing's k industry in the t year, and $X_{ik,t}$ is the k industry's proportion of remaining 12 cities in the t year.

The main sources of these data are Chinese City Statistical Yearbook, Beijing Statistical Yearbook, Tianjin Statistical Yearbook and Statistical Yearbook of Hebei Province from 2005 to 2014. Figure 4 reflects the change trend of the 'benchmark synergy' degree of smoke (powder) dust emissions which is the proxy variable of pollution emission reduction, the degree of industrial structure isomorphism and the strength of economic relation from 2004 to 2013.

In view of the pollution reduction 'benchmark synergy' degree of remaining 12 cities, inter-regional pollution treatment is well collaborated in 2007 - 2010 period, whereas the cooperation between Beijing and Langfang and Hengshui is poor in the period of 2004 - 2006 and there is poor collaboration between Beijing and Hengshui in the period of 2011 - 2013. As the two largest and most populous city in Beijing-Tianjin-Hebei Region, Tianjin's pollution reduction 'benchmark synergy' degree with Beijing has been maintained at a high level. From the point of view of economic relation, the result suggests that the economic relation between Beijing and remaining cities has a trend of overall enhancement shows in the gravity model without standardization. Among them, as the two municipalities of China, the economic relation between Beijing and Tianjin is far more than other cities, while the strong economic relation with Beijing of Langfang city is mainly due to the close distance. From the perspective of the industrial structure isomorphism with Beijing, the industrial structure between Tianjin and Beijing has been increasingly correlated since 2008. In Hebei Province, Qinhuangdao has the highest proportion of the tertiary industry, so this city also has the highest degree of industrial structure isomorphism with Beijing.

(a) Pollution Reduction 'Benchmark Synergy'

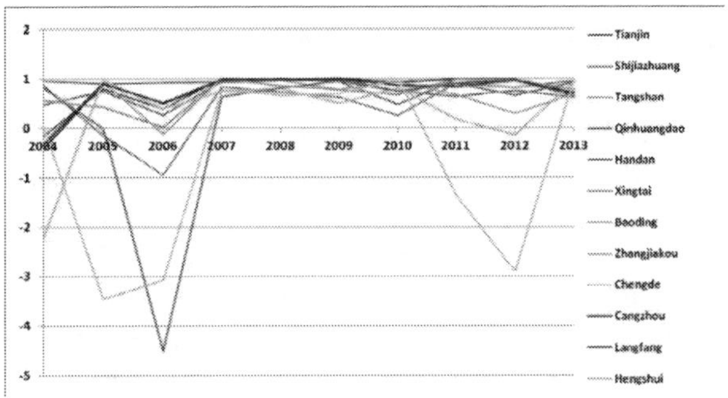

(b) Economic Relation

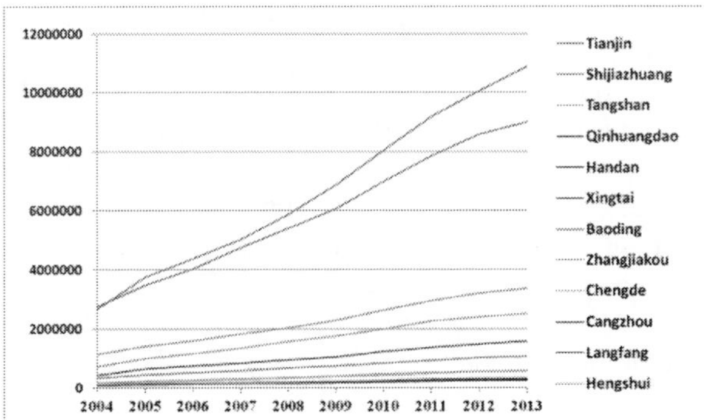

(c) Industrial Structure Isomorphism

Figure 4. The Change Trend of Related Indices

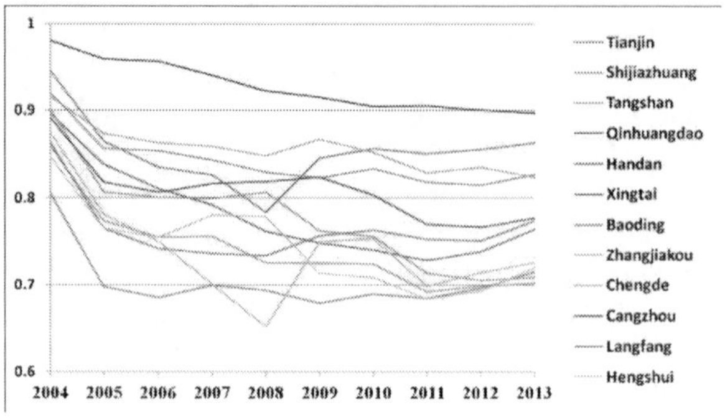

3.2 Model Setting and Results of the Empirical Test

In this paper, explained variable of the model is $Corr_{bi,t}$ the smoke (powder) dust emissions' reduction synergy degree, and explanatory variables are economic relation index $R_{bi,t}$ and industrial structure isomorphism index $H_{bi,t}$. In addition, the model also introduces three virtual exogenous variables that may affect the synergy effect:

$Neighbour_{bi}$ reflects whether the city borders Beijing. On the one hand, the external influence caused by pollution is more obvious among the adjacent regions. On the other hand, it is easier for pollution reduction synergy to take effect for the border area is close. Therefore, the model assumes that the Tianjin, Chengde, Zhangjiakou, Baoding and Langfang 5 cities, which border Beijing, are superior to other cities in the pollution reduction synergy effect.

$Center_{bi}$ reflects whether the city is the central city in Beijing-Tianjin-Hebei Region. Central cities tend to mean more population and economic aggregation. Lu Ming (2014) proposes that the increase of population and economic activity concentration may be beneficial to reduce the emission intensity of the unit economic activities. This model chooses Tianjin and Shijiazhuang as central cities, and assumes that these two cities are stronger than other non-central cities in pollution reduction 'benchmark synergy'.

$Province_{bi}$ reflects whether the city is in the same provincial administrative divisions. The government plays an important role in promoting China's energy conservation and emission reduction. Among them, Wu Liang (2012) thinks that the provincial government plays a more and more important role in the formulation of environmental policy, the pollution and emission reduction regulation and so on. Therefore, the model assumes that the pollution reduction 'benchmark synergy' with Beijing for 11 prefecture level cities in Hebei Province is stronger than Tianjin.

Based on the above ideas, the model is set as follow:

$$Corr_{bi,t} = \beta_0 + \beta_1 R_{bi,t} + \beta_2 H_{bi,t} + \beta_3 H_{bi,t}^{\ 2} + \delta exogenous_{bi} + \varepsilon_{bi,t} \quad (6)$$

$exogenous_{bi}$ is the exogenous variable in the model, mainly including $Neighbour_{bi}$, $Center_{bi}$ and $Province_{bi}$ three dummy variables. $\varepsilon_{bi,t}$ is the random error term. At the same time, the model introduces the quadratic terms of industrial structure isomorphism which can reflect the 'inverted U' type relationship between the degree of industrial structure isomorphism and pollution reduction 'benchmark synergy'. According to the results of theoretical model deduction, there would be $\beta_1 > 0$, $\beta_2 > 0$ and $\beta_3 < 0$ in the formula (6).

In this paper, we first use the OLS method with no fixed effect and random effect to return the formula (6). The results show that the DW statistic is 1.60 and the White statistic is 7.98, indicating that there are serial correlation and heteroscedasticity which suggests that GLS method could be used to improve the estimation of efficiency. Because the section of the model is more than the period of time, the SUR Period method is chosen to estimate, and Table 1 is the regression results. The Model (1) in Table 1 reflects the influence to the 'benchmark synergy' from the degree of economic relation and the degree of industrial structure isomorphism.

The results show that the enhancement in economic relation with Beijing for remaining 12 cities has a positive effect on inter-regional pollution reduction 'benchmark synergy' in the 1% significant level. It means that with deepening economic integration and economic relation in Beijing-Tianjin-Hebei Region, the synchronization and consistency will gradually increase for the related area with Beijing in smoke (powder) dust emissions' reduction. The coefficient of the industrial structure isomorphism of the first term is positive and the two terms are negative, which confirms the existence of the 'inverted U' type relationship between the industrial structure isomorphism and pollution reduction 'benchmark synergy'. That is to say, the over homogeneous and over foreignizing industrial structure is helpless for the pollution reduction 'benchmark synergy' effect to take. Enhancing cross-regional industry correlation and forming complementary advantages of industrial structure are also the key to solve the problem.

The estimated results from Model (2) to Model (4) show that the 'benchmark synergy' effect of smoke (powder) dust emissions' reduction generated by the enhanced economic relation is still significant after adding exogenous variables one by one. And the 'inverted U' type relationship between the industrial structure isomorphism and the pollution reduction 'benchmark synergy' is still maintained, which proves the stability of the research conclusion. The Model (5) comprehensively reflects the regression results with all variables. The results show that, in addition to the existing relationships between the three core variables, the theoretical assumptions of the three dummy exogenous variables are all established: the pollution reduction 'benchmark synergy' effect occurs in the adjacent regions of Beijing in the 5% significant level.

Due to the fact that the central cities may be more similar in the mode of economic development and environmental protection measures, the smoke (powder) dust emissions' reduction 'benchmark synergy' with Beijing of Tianjin and Shijiazhuang two cities differs from other non-central cities at the 10% level of significance; 11 prefecture level cities of Hebei Province have a higher smoke (powder) dust emissions' reduction synergy effect than Tianjin in the 1% significant level, so it can be shown that the same provinces are easier to realize the synergy of pollution reduction.

4. CONCLUSION

Theoretical model and empirical results revel following conclusions:

(1) The 'benchmark synergy' effect of pollution reduction has a monotonically increasing relation with economic relation. The increase of regional economic relations is conducive to the implementation of coordinated inter-regional pollution reduction.

(2) Pollution reduction 'benchmark synergy' and industrial structure isomorphism have 'inverted U' type relation. Regions should maintain appropriate industrial structure isomorphism level, thus maintaining high industrial correlation within the whole region, which is conducive to pollution reduction 'benchmark synergy' effect to play.

(3) The 'benchmark synergy' effect of pollution reduction is easier to emerge between adjacent areas on the administrative divisions or between central cities. And there is a difference between Tianjin and Hebei Province in pollution reduction synergy effect, which indicates that the same provinces are more likely to be in 'lock step' on pollution reduction.

Table 1. Regression Results

Explanatory Variables	Model (1)	Model (2)	Model (3)	Model (4)	Model (5)
Constant term	-3.294*** (-10.07)	-2.313*** (-7.08)	-2.403*** (-8.74)	-1.941*** (-7.31)	-1.990*** (-6.75)
R	0.265*** (21.54)	0.268*** (25.20)	0.266*** (26.19)	0.262*** (27.66)	0.265*** (27.25)
H	8.044*** (9.20)	5.668*** (6.44)	5.916*** (7.83)	5.368*** (7.45)	5.335*** (6.81)
H^2	-4.056*** (6.95)	-2.526*** (-4.26)	-2.777*** (-5.45)	-2.504*** (-5.16)	-2.405*** (-4.56)
Neighbour		0.140 (3.74)			0.094** (2.33)
Center			0.020*** (0.32)		0.127*** (1.75)
Province				0.240*** (3.06)	0.289*** (2.94)
R^2	0.861	0.877	0.890	0.896	0.893
DW	1.97	1.97	1.99	1.99	1.98
$F-statistic$	238.52	204.21	232.91	248.22	156.88
Obs.	120	120	120	120	120

Note: The ***, ** and * in the table respectively indicate that they could pass the significance test at the level of 1%, 5% and 10%

Therefore, in order to enhance pollution reduction synergy of the Beijing-Tianjin-Hebei Region, and to achieve the coordinated development of economy and environment in the whole region, economic relation between cities within the whole region should be firstly strengthened. Cities should be inseparable with close relationship. At the same time, the industrial layout of Beijing, Tianjin and Hebei Province should be rationally planned to form a complementary industrial structure. It is necessary for cities within the region to cooperate on inter-regional industrial docking, and enhance industry correlation. In this process, on the one hand, an overall planning for Beijing-Tianjin-Hebei Region's economic development as well as industrial layout is important. On the other hand, for the fact that Beijing is a relatively developed areas of the economy and Hebei Province is characteristic of heavy industrialization structure with relatively less developed economic, Beijing should bring the 'benchmark' area's economic function into full play, which means providing more economic support to Hebei Province such as in industrial transfer so as to alleviate the pollution reduction pressure of Hebei Province.

REFERENCES

- Diamantoudi, E. and Sartzetakis, E. S. (2006). Stable international environmental agreements: An analytical approach. *Journal of Public Economic Theory, 8(2),* 247-263.
- Rypdal, K., Rive, N., Åström, S., Karvosenoja, N., Aunan, K., and Bak, J. L., et al. (2007). Nordic air quality co-benefits from european post-2012 climate policies. *Energy Policy, 35(12).*
- Li Xuesong and Sun Bowen (2014). Economic property and policy development of atmosphere pollution governance: An analytical framework. *Reform, 4,* 17-25.
- Zhang Yanbo, Tong Linjie, and Meng Weidong (2015). Research on ecological governance perspective regional government collaboration in Beijing-Tianjin-Hebei Region. *Economy and Management, 29(3),* 23-26.
- Huang Binhuan, Yang Haobo, and Yao Maohua. (2015). Reconstruction of powers, production of society and collaborative governance of ecological environment. *China Population, Resources and Environment, 25(2),* 105-110.
- Cui Jing and Sun Wei (2014). Intergovernmental jurisdiction division in regional air pollution governance from the perspective of collaboration governance. *Chinese Public Administration (9).*
- Yu Dianfans, Gan Chunhui, and Zheng Ruogu (2011). Industrial linkages in China – An empirical research based on structural decomposition technique of input-output model. *China Industrial Economics, 11,* 5-15.
- Cerqueira, P. A. and Martins, R. (2009). Measuring the determinants of business cycle synchronization using a panel approach. *Economics Letters, 102(2),* 106-108.
- Lu Ming and Feng Hao (2014). Agglomeration and emission reduction: An empirical study on the impact of urban scale gap on the intensity of industrial pollution. *The Journal of World Economy, 7,* 86-114.
- Wu Liang, Ma L, and Qi Ye (2012). An analysis of policymaking process of provincial environmental policy: Case study of the ecological compensation policy for conservation of water and soil in Shannxi and Yunnan. *China Population, Resources and Environment, 22(3),* 87-92.

Session II: Measures of Success

A Research on the Reason of Sinusoidal Fluctuation Process of the Growth Rate of Chinese Housing Price Index During 2012 to 2015

Liu Siyuan
Postgraduate Student, Tianjin University of Finance & Economics

Cong Yi
Professor, Tianjin University of Finance & Economics

1. INTRODUCTION[1]

From 2012 to 2015, the growth rate of China's one-hundred-cities housing price index has experienced an sinusoidal fluctuation process, Simultaneously, the Shanghai and Shenzhen 300 index experienced an inverted U-shaped changing process, the exchange rate of CNY relative to USD experienced an U-shaped process, either - appreciate and then devaluate, the average growth rate of M2 was 13.43%, Moreover, China's regulation policy of housing market has changed from "trade restricting" to "storage clearance". By means of correlation analysis this paper aims to find the explanatory variables of the sinusoidal process of China's one-hundred-cities housing price index. In our opinion, housing price is one of the generalized asset prices in China. So, China's housing price can be affected mainly by three kinds of factors. Firstly, it can be affected by the policies and regulations. Secondly, it can be influenced by variables of financial market. Thirdly, it can be restricted by the supply and demand forces in housing market. Moreover, the situations between the first-tier cities, the second tier cities and the third-tier cities are quite different, so the policies should be published respectively and differently between the first-tier cities, the second tier cities and the third-tier cities.

2. REVIEW OF CHINESE HOUSEING MARKET

2.1 Changing Process of Chinese One-Hundred-Cities Housing Price Index

Chinese one-hundred-cities housing price index increased during July 2012 to March 2014. While from April 2014 to March 2015, the index fell down, and then it rose up. Moreover, it was easily to observe a sinusoidal process if we watched on the year-on-year growth rate of Chinese housing price index of one hundred cities. We could find the trend of the data by using the method of Hodrick-Prescott Filter. Figure1 shows the trend of the growth rate of China's one-hundred-cities housing price index among China, China's first-tier cities , the second-tier cities and the third-tier cities.(PI represent the price index among China, FPI represent the index of first-tier cities, SPI represent the price index of second-tier cities, and TPI represent the index among third-tier cities of China.) It is obviously that the trend of all of the four growth rates of housing price index declined, the situation in third-tier cities illustrated the falling trend straightly, and in the first-tier cities, the growth rate was relatively stable. It was worth noting that there was only seven months when FPI became negative, there was twenty months when SPI turned to negative, and TPI was negative in twenty-six months. By November 2015, TPI was kept under zero for seventeen months and SPI had continued under zero for fifteen months.

2.2 Changing Process of Government's Regulations and Rules

The attitude toward housing construction in Beijing has experienced several transitions from the establishment of housing market in China. From 1990 to 2001, Beijing encouraged the development of housing market and housing construction. From 2002 to 2005, some policies in support of housing market development was canceled and the banking system was tighten-up accordingly. After 2005, regulation policies turned tighter. *Notice on Strengthening the Credit Management of Commercial Real Estate* was published in 2007, which posed a higher requirement on the management of loan providing of development of real estate, besides, it requires regulating land reserving loans and improving down payments for second-home purchases and lending rates. For the financial crisis of the world broke up in 2008, China's economy and society have suffered a serious impact from the world, as a important engine of China's economic development, the real estate market ushered in a brief and loose policies, including reducing the down payments for second-home purchases and lending rates and improving the loan property which

[1] *The work is supported by the project of graduate research fund (2014TCS01) 'A research on the trend of the Chinese real estate market development and related regulation and control of the government' supported by Tianjin University of Finance and Economics.*

housing buyers could apply. But in 2009, some problems becomes more serious: a great number of housing buyers purchased houses neither for living nor for investment, but for speculate; most state governments took housing development as a measure to improve GDP growth rate and revenue; housing price became so high that normal families realized that it was a huge burden to buy a house to live in. Beijing decided to restrain unreasonable demand in real estate market, including housing market. Relevant departments decided to improve business tax exemption period of personal housing transfer from two years to five years.

From 2010, a lot of policies were published to restrict speculate housing demand, and to avoid financial risks in real estate market. On January 10th, 2010, General Office of the State Council asked to strengthen and improve the regulation of real estate market, to keep the expectation of real estate market stable, and to promote real estate market's development with a stable and healthy character. On April 11th, China Banking Regulatory Commission (CBRC) asked banking system not to provide loans for those who purchased real estate for investment or speculate. The Standing Committee of the State Council set that the proportion of down payments for second-home purchases can not be under fifty percent on April 14th. Then State Council required banking system to stop providing loans for third-home purchases in many regions with excessive high housing price on April 17th. On October 20th, Ministry of Housing and Urban-Rural Development published management regulation explaining that deposit rate of Personal Housing Accumulation was rose up by 0.2 percent to 1.91 percent, loan rate of Personal Housing Accumulation of five years or less was rose up by 0.17 percent to 3.5 percent, and loan rate of Personal Housing Accumulation for more than five years rose up by 0.18 percent to 4.05 percent. Ministry of Housing and Urban-Rural Development, Ministry of Finance, PBC, CBRC regulated on November 3rd that the proportion of down payments for first-home (more than 90 square meters) purchases using Personal Housing Accumulation loans could not be under thirty percent. The proportion of down payments for second-home purchases using Personal Housing Accumulation loans could not be under fifty percent, loan rate could not less than 1.1 points of which of Personal Housing Accumulation; banking system could not provide Personal Housing Accumulation loans for those who wants to buy three or more houses. In order to control demand of real estate from foreign persons, enterprises and organizations, Ministry of Housing and Urban-Rural Development, and State Administration of Foreign Exchange

Figure 1. Trend of the growth of Chinese One-Hundred-Cities Price Index Drawn Using Hodrick-Prescott Filter

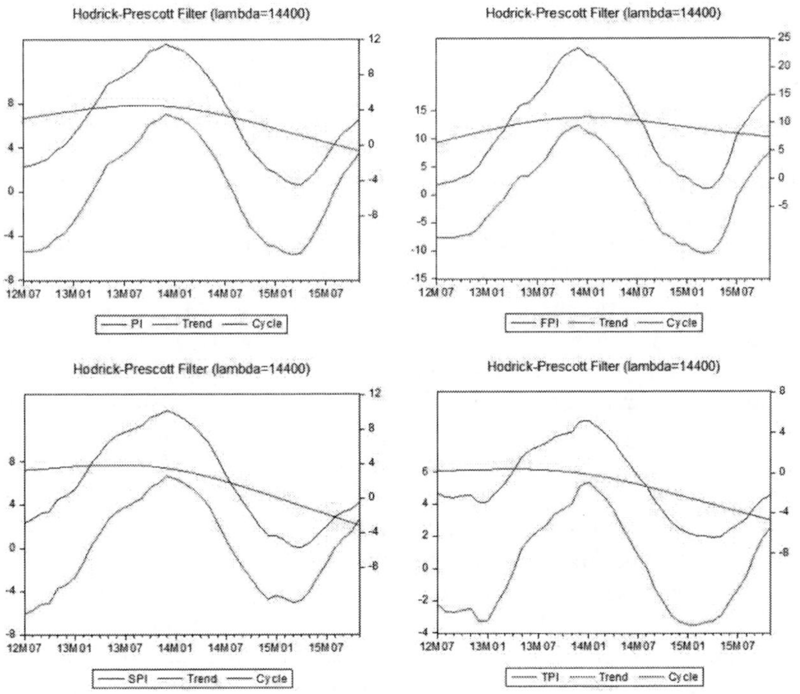

Source: www.windin.com

(SAFE) published a regulation rule on November 15th which said that foreign people could only buy one house for living in, and foreign institutions which established branches and representative offices could buy non-residential housing only in the cities where it registered.

Chinese former Premier Wen Jiabao chaired a State Council executive meeting on January 26th. 2011 and determined six main policy instruments: firstly, improve supply for public rental housing; secondly, those who resells second-hand houses in five years after he or she purchased taxed it according to total housing price; thirdly, the proportion of down payments for second-home purchases was rose up to 60 percent; fourthly, usage rights would be taken back if the land do not start in two years; fifthly, paused the rights to buy third houses in provincial capital cities; finally, state governments who could not complete controlling targets of housing price and those who spread false information would be blamed or punished.

New Five Regulatory Policies on the Real Estate was published on February 22th, 2013, which especially regulated the proportion of down payments becoming higher, and indemnificatory housing development was encouraged. Especially for those who resold second hand houses would be levied according to twenty percent of the profit.

On April 25th, 2014, PBC announced to reduce the RMB deposit reserve ratio in Xians' rural commercial banks by 2 percent and to reduce which in Xians' rural cooperative banks by 0.5 percent. On September 30th, PBC published new regulations of loans: firstly, the proportion of down payments for first-home purchases declined to thirty percent and the lower limit for personal housing loan rate could be 0.7 times of the benchmark; secondly, families which had already repaid the housing loans purchasing a new improving house could enjoy the preferential policy for the first houses; thirdly, families who already had two houses could apply loans from commercial banks if the loans were paid off before.

PBC announced on December 21th to reduced benchmark interest rate of loans by 0.4 percent to 5.6 percent, this was the first time PBC reduced the benchmark interest rate in two years. Moreover, there were five times that PBC reduced the benchmark interest rate in 2015: On March 1st, PBC reduced the benchmark interest rate of loans and deposit respectively by 0.25 percent to 5.35 and 2.5 percent; On May 11th, benchmark interest rates were reduced by 0.25 percent, and then, the rates were cut down respectively by 0.25 percent on June 28th and August 26th. On November 24th, benchmark interest rates were reduced by 0.25 percent, but there were no changes in loan rates and deposit rates of Personal Housing Accumulation Fund. Correspondingly, deposit reserve ratio was cut down five times for specific banks and other financial institutions representatively in February 4th (0.5 percent to all, excess 0.5 percent for commercial banks who support minor enterprise and some rural commercial banks, and excess 4 percent for Agricultural Development Bank of China), April 20th (1 percent to all, and excess 1 percent to rural banks and financial institutions), September 6th (0.5 percent to all, excess 0.5 percent to rural banks, rural cooperative banks and rural financial institutions, and excess 3 percent to finance lease companies and auto financial enterprises), and October 24th (0.5 percent to all). Reasons of policies about reducing benchmark interest rates and deposit reserve ratios are comprehensive, but it is a significant factor which operates up China's real estate market including housing market.

On March 30th, 2015, *Notice on the relating issues of the policies of individual housing loans* was published, according to this notice, families which hadn't paid off loans yet could enjoy the down payments proportion of forty percent when they bought the second improving houses; families could enjoy the down payments proportion of twenty percent when they buy first houses using Personal Accumulation Fund loans. Besides, families which had already paid off previous loans yet could enjoy a down payments proportion of thirty percent when they bought a second improving house with the help of Personal Accumulation Fund loans. Meanwhile, *Notice on the policy adjustment of business tax of individual housing transfer* was published, in this notice, business tax exemption period of personal housing transfer was reduced from five years to two years. On December 18th to 20th, "reduce the stock" was proposed in the Central Economic Working Conference of China held in Beijing, this concept meant that the rate of household population urbanization should be improved by transfer persons from migrant worker to citizens, state governments should encourage the migrants to take up residence in local area, besides, housing lease market and public lease houses should be established, promoted and developed. Policies which did not match current situation and circumstance should be abolished. Finally, it should be encouraged for housing constructing enterprises to combine and discount.

Figure 2. Changing process of the growth of Chinese One-Hundred-Cities Price Index

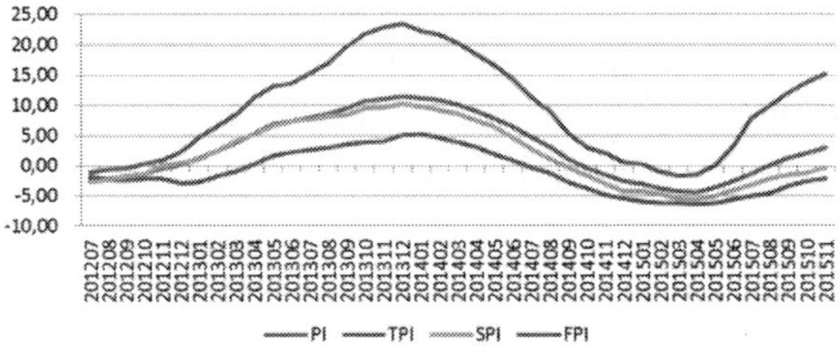

Source: www.windin.com

Figure 3. Changing Process of Relative Variables in Supply Aspect in Housing Market

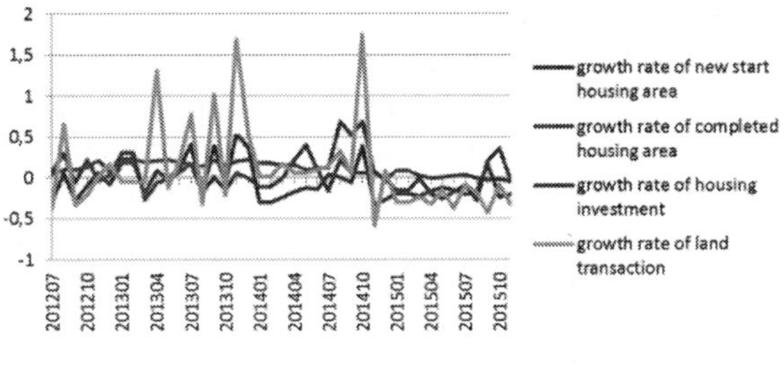

Source: China Economic and social development statistics database

Now, let us review the changing process of the growth of China's one-hundred-cities housing price index referring to the policies above and from figure2,we can come to a conclusion that the policies which aimed at restricting the demand volume of housing market haven't played a significant role, while the policies encouraging the housing demand led to the improvement of the growth rate of housing price index during March 2015 to November 2015. Moreover, it can be found that there are some difference between the policies during 2009 to 2013 and which during 2014 to 2015. The former policies aims to restrict demand quantity of houses by many limiting measures toward housing buyers, and the later policies aims to encourage living demand by encouraging housing loans especially Personal Accumulation Fund loans. We can find that the later policies seem more targeted and discriminating comparing with the former policies. But there is still a question that why the housing price index's growth rate declined from January 2014 to March 2015, with few policies published. This question will be researched later in this paper.

2.3 Changing Process of other Relative Variables in China's Housing Market

Concerning about supply aspect of housing market in China, there are four main variables which could manifest current supply situation of China's housing market. Referring to figure3, the growth rate of new start housing area, completed housing area, housing investment, land transaction have got down in the past three and a half years. This phenomenon shows that enterprises engaging in housing construction are not full of enthusiasm any more, though the price index rose up during April 2015 to November 2015. So, we can find it obviously in figure4 that the growth rate of housing pending sales experienced a decline process from December 2013 to November 2015.

Figure 4. Changing Process of Growth Rate of Pending Sales of Houses

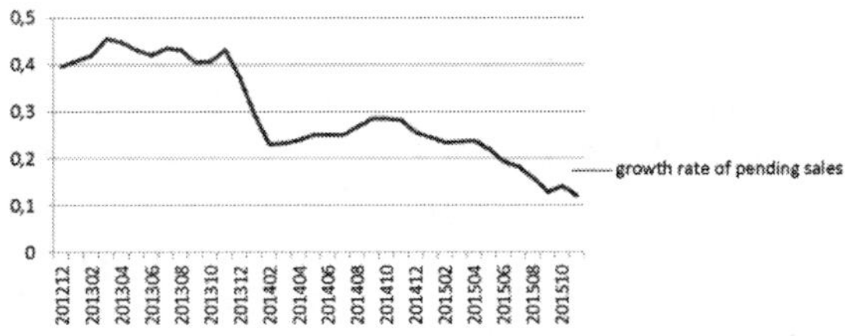

Source: China Economic and social development statistics database

It may be good news. Arguments about the disadvantage of excess housing supply can often be heard in the past, and we can see a lot of dark windows at evening in a huge number of communities, especially in the second-tier cities and the third-tier cities. If the growth rate of pending sales continues to decline, Beijing will be more self-confident in "reduce the stock". What may be confused is why housing constructing enterprises do not increase investment in housing construction as before even the benchmark interest rate has fallen in 2015 and why the enthusiasm of housing construction does not rise up during the process of the improvement of housing price index.

The answer of the first question may be that the housing constructing enterprises do not have enough funds to increase housing investment, or that the housing constructing enterprises hold little confidence in housing investment in the future, especially in the second-tier and third-tier cities of in China. This partly because a huge number of housing constructing enterprises still have a lot of liabilities from loans or still have lending funds haven't been paid off yet. For the reason above, though the policies encourage housing investment, a huge number of medium and small sized enterprises even some big sized enterprises can not apply new loans for the restrict of credit level and funding chain of itself or loans and risk controlling policies of commercial banks. On the other hand, though the policies after 2014, especially in 2015 encourage housing investment objectively, but some of the policies are not published only for real estate market. And the attitude of government is restricting high growth rate of housing price. For this reason, enterprises should consider carefully and make decisions cautiously when they plan to improve housing investment.

Concerning about demand aspect, we can illustrate current Chinese housing market with three main variables: growth rate of personal mortgage loans, growth rate of housing sales and benchmark lending rates. Figure5 shows the changing process of these variables. It can be acknowledged that the decrease of the benchmark lending rate led to the increasing of growth of personal mortgage loans and housing sales for the process of the decrease of benchmark lending rate start earlier, and referring to figure2, the increasing of growth of personal mortgage loans and housing sales is the reason of improvement of the growth of housing price index but not the result for the trend of decline of the former starts during the growing process of the later. For loans support investment or speculate housing demand are not permitted from 2010, personal mortgage loans can be considered as a variable indicating housing demand for living.

Figure 5. Changing Process of Relative Variables in Demand Aspect in Housing Market

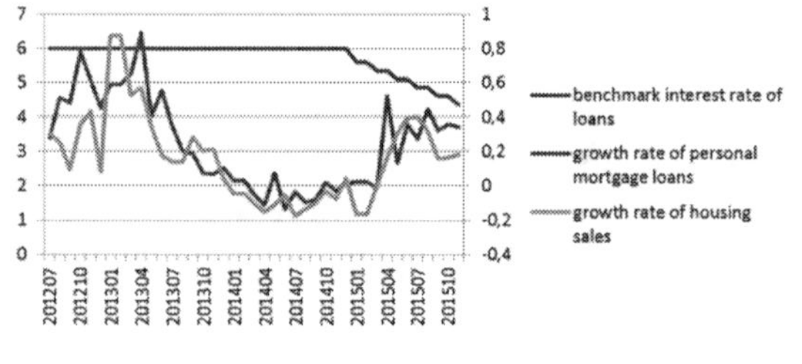

Source: China Economic and social development statistics database

In figure 5, growth rate of personal mortgage loans and housing sales are significantly correlated positively, it means that housing demand for living is a important force in China's housing market, and it is sensitive toward encouraging policies, and it can be trusted that Beijing could do a good job in "reduce the stock" based on a serious of encouraging policies toward living housing demand.

3. FINANCIAL AND ECONOMIC INFLUENCING FACTORS OF THE SINUSOIDAL PROCESS

3.1 The Factors which Led to the Sinusoidal Process of the Growth of China's Housing Price Index

After 2008, the world is bearing the pain of global financial crash which broke out in USA and then impacted all around the world. China's economy is facing a downward pressure, and there are some characters emerged though the Chinese economy is running stable in general. Two of the most obvious characters are that the GDP growth rate comes to 6.9 percent in 2015 in general, and the stock market crash in 2015. In figure6, Shanghai and Shenzhen 300 Index rose up rapidly from July 2014 to June 2015 and then fell down quickly. SSI (Shanghai and Shenzhen 100 index) was stabilized at 2400 when growth rates of housing price indexes was positive from December 2012 to September 2014, and SSI increased when growth rates of housing price indexes became negative in general. Though SSI fell down from June 2015, it still much higher than 2400 in November 2015. This means that China's stock market and housing market substitute mutually.

Figure6. Changing Process of SSI, PI, FPI, SPI and TPI

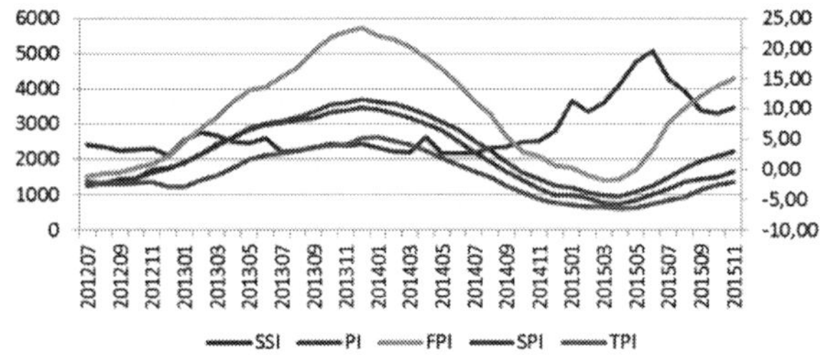

Source: China Economic and social development statistics database and http://www.10jqka.com.cn/

Housing price is often considered as one of the asset prices in China, depreciation process of RMB during July 2015 to November 2015 great impacted on China, so does its impact on China's housing market. In figure7, we can find that exchange rate in direct quotation rose up company with the increasing of the growth rate of housing price index. It means that people would like to buy houses when RMB depreciates. For the reason above, it can be judged that a house is still an important hedging and investment asset in China. But Governments must consider the basic living function of houses and income level of average persons in China. The housing price could not increase so fast that persons in working class can not afford. Relevant policies may be published later to ensure that the housing price won't be so high, and Beijing is endeavoring in extending the channel of investment.

Currency supply is another important factor which can take effect toward housing price index.China's M2 growth rate kept stable in general at the average of 13.43 percent, from July 2012 to November 2015. But the M2 growth rate of RMB appeared a U-shaped changing process: it dropped slowly to 10 percent and then increased to 14 percent, just as what figure8 shows.

Figure 8. M2 Growth Rate of RMB

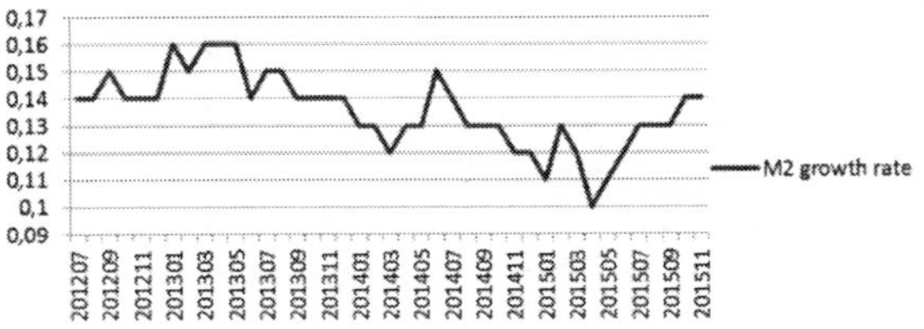

Source: http://data.eastmoney.com/cjsj/hbgyl.html

Figure 7. Changing Process of Exchange Rate and Other Relevant Variables

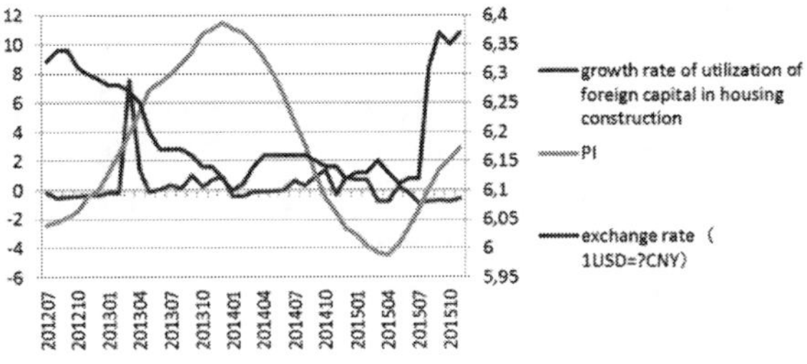

Source: China Economic and social development statistics database

3.2 Correlation Analysis and the Conclusions

According to the analysis above, we choose four main explanatory variables including growth rate of M2 (be represented by GM), Shanghai and Shenzhen 300 Index(be represented by SSI), exchange rate between RMB and USD in direct quotation(be represented by EXR), and benchmark interest rate of loans (be represented by BIL) and research the correlative relationship between FPI, SPI, TPI, GM, SSI, EXR, and BIL. Table1 is the correlation matrix of these variables.

Firstly, correlative relationships between FPI, SPI, and TPI is significant and great, which means there are some demonstrative effects between housing prices in the first-tier, the second-tier and the third-tier cities. It means that housing price in second-tier cities and third-tier cities may rise up when the housing price in first-tier cities grows up.

Secondly, SPI and TPI have positive correlation with GM, but the positive correlative relationship between FPI and GM is neither sufficiently significant nor large. It partly because the changing trend of FPI is stable referring to figure1. This phenomenon shows that housing price in first-tier cities is well supported. Characters of value preservation and appreciation of houses in first-tier cities are so strong that people tend to

Table 1. Spearman Correlation Matrix of Chosen Variables

	FPI	SPI	TPI	GM	SSI	EXR	BIL
FPI	1						
SPI	0.897**	1					
TPI	0.881**	0.979**	1				
GM	0.277	0.465**	0.435**	1			
SSI	-0.358*	-0.613**	-0.653**	-0.552**	1		
EXR	-0.233	-0.234	-0.183	0.466**	-0.104	1	
BIL	0.109						
-0.936**	0.490**						
-0.886**	0.492**						
-0.901**	0.335**	-0.780**	-0.280	1			

a.** represent the correlative relationship is significant on 0.01 percent level
b.* represent the correlative relationship is significant on 0.05 percent level
c. In correlative coefficients, numbers in the first line represent the situation in the period from July 2012 to November 2015; numbers in the second line represent the situation in the period from January 2015 to November 2015.

regard them as one of the best asset. For this reason, we can observe that the correlative relationship between FPI and SSI is much smaller than which between SPI and SSI, or TPI and SSI. While the situation of SPI and TPI means that increasing currency supply can promote the growth of housing prices. On the one hand, all of the assets marked by CNY will increase if money supply of RMB grows up. On the other hand, People tend to buy houses which are good assets for value preservation during the period of inflation.

Thirdly, FPI,SPI and TPI have negative correlation with SSI, except the correlative coefficient between FPI and SSI is -0.358, the correlative coefficient between SPI and SSI reaches -0.613, moreover, correlative coefficient between TPI and SSI even reaches -0.653. It can be acknowledged for this situation that houses are great alternatives for stocks. And the housing values from high to low is first-tier cities, second-tier cities and third-tier cities. As we can see in figure6, the alternative relationship between growth of housing price indexes and SSI become more obvious after October 2014.

Fourthly, exchange rate doesn't have great effect on FPI, SPI and TPI. On the one hand, Beijing implements exchange controls, so people can not purchase excess foreign assets when RMB is depreciating. On the other hand, foreign capital doesn't account for an important proportion: actual utilization of foreign capital in real estate development is steady under 10 billion CNY, while the average amount of housing investment was 514.4 billion CNY from July 2012 to November 2015. Meanwhile, according to the policy published by SAFE on November 15th, 2010, foreign people only can buy one house for living in, and foreign institutions which established branches and representative offices can buy non-residential housing only in the cities where it registered.

Besides, SPI and TPI have positive correlation with BIL, respectively the correlative coefficient are 4.9 and 4.92. Considering BIF stays at 6 percent before 2015 according to figure5, we calculate the correlative coefficient between FPI, SPI, TPI and BIF. The outcome is significant, large enough and negative, which means that personal housing loans can obviously gain and then promote the growth of housing price indexes while benchmark interest rate of loans goes down, just like what is illustrated in figure5.

Finally, we can find other meaningful relationships between the explanatory variables, SSI has negative correlation with GM, and GM has positive correlation with SPI and TPI. It seems that when housing price goes down, capital will flow toward stock market, which validates alternative relationship between houses and stocks.

4. SUMMARY AND CONCLUSIONS

In our opinion, housing price is one of the generalized asset prices in China. So, China's housing price can be affected mainly by three kinds of factors. Firstly, it can be affected by the policies and regulations. Secondly, it can be influenced by variables of financial market. Thirdly, it can be restricted by the supply and demand forces in housing market.

From the policies aspect, policies which aimed at restricting the demand volume of housing market haven't played a significant role, policies encouraging the housing demand led to the improvement of the growth rate of housing price index.

From the financial and economic aspect, in general, currency supply has positive correlation with housing price. Stock price has negative correlation with housing price. Benchmark interest rate has negative correlation with housing price when it decreases. Besides, it can be judged that a house is still an important hedging and investment asset in China. But the situations between the first-tier cities, the second tier cities and the third-tier cities are quite different as it illustrated above. So the policies should be published respectively and differently between the first-tier cities, the second tier cities and the third-tier cities.

From housing supply aspect, enterprises engaging in housing construction are not full of enthusiasm any more. While, housing demand is the reason of improvement of the growth of housing price index. Housing demand for living and improvement is an important force in China's housing market, and it is sensitive toward encouraging policies. So we keep an optimistic attitude toward China's housing market in the future.

REFERENCES

- Xiao Zhengyan and Peng Bo (2011). House Prices and Monetary Policy Rules in China [J]. *Statistical Research,11,* 40-49
- Jiao Jiwen and Chen Wentian (2012). A Research on the Period and Relevance of real estate market of China Based on HP Filter [J]. *Construction Economy, 8,* 79-83.

- An Hui, WANG Li-ting, and Gu Yu (2014). A Research on Influencing Factors of Housing Prices on Both the Long-term Trends and Short-term Fluctuations: Based on Panel Data of 35 Large and Medium-sized Cities in China [J]. *Forecasting, 2,* 13-19.
- Cong Yi (2015). New Policies of Real Estate, Not the Simple Bailout Policies [J]. *Social Outlook,5,* 55-56.
- Yu Yangmin (2013). The Research for the Political of the Periodic Fluctuation in Chinese Real Estate: Theory and Empirical Analysis [D]. East China Normal University.

Intellectual Property Protection and GVCs Position A Study Using Koopman Index

Zhenzeng YANG
Ph.D., Assistant Professor, Tianjin University of Finance and Economics, China

1. INTRODUCTION

Differences in the level of intellectual property rights (IPR) protection have an important influence on global trade and foreign direct investment patterns. Previous literature shows that lower intellectual property protection in developing countries hinders the multinational direct investment in these countries. But in the background of global value chains (GVCs), a large number of international direct investment flows to developing countries, and a considerable portion of these direct investment flows into the high-tech sectors, which directly contributed to the rise in the proportion of high-tech exports in developing countries. On the other hand, a noteworthy phenomenon is that the domestic value-added in high-tech exports of developing countries remains low, which indicates that multinational firms only put those production processes with low value-added and low technology into developing host countries. I focus how IPR protection affects multinational corporations' pattern of international production network, and investigate whether stronger intellectual property protection can attract MNCs locate more high value-added production processes into host countries. In other words, whether highly intellectual property protected countries have higher global value chain status.

2. PREVIOUS RESEARCH

IPR protection in host countries has an important impact on the direct investment of transnational corporations. Higher IPR protection can prevent the entrance of imitators, lower the speed of the host local enterprises to imitate, and stimulate foreign direct investment inflow, and spur local production scale in host countries (Helpman 1993, Lai 1998, Yang & Maskus, 2001, Markusen 2001, Park & Lippoldt 2005, Mathewa & Mukherjeea 2008, Awokuse & Gu 2010, Branstetter et al. 2011, etc.). Not only direct investment, host countries with higher IPR protection can spur exports and technology transfer from developed countries (Ghosh & Ishikawa 2013). Hence higher IPR protection in host country can lead to industrial expansion and higher creative ability (Smith 2001, Branstetter et al. 2006, Ivus 2010, Branstetter & Saggi 2011, Tanaka & Iwaisako 2013). The impact of IPR protection on FDI varies between industries (Nunnenkump & Spatz 2004, Bilir 2011).

Multinational corporations in low technology industries such as clothing, textiles, mining and processing, electronic assembly manufacturing are less sensitive to local IPR protection when they invest abroad. But market size, operating costs and other factors matter (Kalande 2002). Firms in industries with high R&D expenditure, low productive complexity and low imitation cost, are more sensitive to local IPR protection. But the impact of IPR protection are unclear to firms in industries with high R&D expenditure and low imitation cost, like bio-pharmaceutical, chemical, software development (Markusen 2001). Nabokin (2014) examines the impact of intellectual property protection for foreign multinational investment decisions using the data of German multinational corporations, and shows that IPR protection matters for extensive margin of German multinational companies overseas direct investment. And this positive impact is higher in industries with higher productive complexity. Lippoldt (2006) also found that strengthening of intellectual property protection helps attract multinational corporations to transfer high technical complexity production to host countries.

In the context of global value chain, multinational corporations put production processes that are more sensitive to IPR protection into countries with higher IPR protection level. This pattern of production will then have influence on host countries global value chain position: countries with higher IPR protection will enjoy higher GVC position. Among previous literature, few provide empirical evidence that IPR protection help to improve a country's GVC position. This study contributes in this field by empirical work using cross-country data.

3. EMPIRICAL STRATEGY

To investigate whether stronger IPR protection leads higher GVC position, I specified the following empirical model:

$$\ln(1+GVC\_position_{ij,t}) = b_0 + b_1\ln(IPR_{i,t}) + b_2\ln(PINT_{j,t}) + b_3\ln(IPR_{i,t})*\ln(PINT_{j,t}) + b_4\ln(KEND_{it}) + b_5\ln(HC_{it}) + b_6\ln(KENT_{it})*\ln(HC_{it}) + b_7\ln(Topen_{i,t}) + e_{it} \quad (1)$$

where the dependent variable $\ln(1+GVC\_position_{ij,t})$ is the Koopman GVC position index in the year of t for individual 'country-industry pair' ij. $\ln(IPR_{i,t})$ presents effective IPR protection level in country i in year of t. $\ln(PINT_{j,t})$, $\ln(KEND_{it})$, $\ln(HC_{it})$ and $\ln(Topen_{i,t})$ present industry patent intensity, capital endowment, human capital index and trade openness respectively. I also specified the following empirical model to finish dynamic panel estimation:

$$\ln(1+GVC\_position_{ij,t}) = b_0 + h\ln(1+GVC\_position_{ij,t-1}) + b_1\ln(IPR_{i,t}) + b_2\ln(PINT_{j,t}) + b_3\ln(IPR_{i,t})*\ln(PINT_{j,t}) + b_4\ln(KEND_{it}) + b_5\ln(HC_{it}) + b_6\ln(KENT_{it})*\ln(HC_{it}) + b_7\ln(Topen_{i,t}) + e_{it} \quad (2)$$

4. DATA AND MEASUREMENT

4.1 GVC Position

Koopman (2010) uses the following method to calculate countries GVC position:

$$GVC\_Position_{ir} = Ln\left(1+\frac{IV_{ir}}{E_{ir}}\right) - Ln\left(1+\frac{FV_{ir}}{E_{ir}}\right)$$

where IV is indirect value added in exports to third country via importer, FV presents foreign value added in export, and E is total export. The data used in this study is drawn from Wang et al. (2013). The dataset used in this paper includes 35 economies (covering most economies in WIOT), 18 manufacturing sectors (from c01 to c17 plus c30), in the period of 2000-2009.

4.2 Effective IPR Protection (IPR)

Effective IPR protection index is calculated using the following method:

Effective IPR protection = GP index * Fraser index

GP index is compiled by Ginarte and Park (1997) and updated by Park (2008). The GP index assigned each country a score between 0 and 1 for each of five elements: the extent of coverage of patent protection, membership in international patent agreements, provisions for protection against loss, enforcement mechanisms, and duration of protection. The GP index was the unweighted sum of these individual scores and ranges from 0 to 5.

The GP index measured the completeness of laws de jure. But effective IPR protection depends on both the formation of IPR law and its enforcement. So I use the method by Hu and Png (2012) to calculate effective IPR protection index by multiplying GP index by fraser index. The fraser index data is drawn from EFW 2015 dataset.

4.3 Industry Patent Intensity (PINT)

For each industry, we measured patent intensity by the ratio of the number of patents granted to an industry to the total value added of the industry. That is:

Patent intensity = Patent granted to industry i / total value-added of industry i

Patent number data is drawn from the website of US Patent and Trademark Office. And data of industrial value-added is drawn from NBER-CES dataset.

4.4 Capital Endowment (KEND)

In the circumstances of global value chain, according to the H-O theory, capital abundant countries will be specialized in producing capital-intensive production processes, which have relatively higher added value. In GVC, these countries usually export core components, which may be export to other countries after further process. So capital endowment index is expected to be positively correlated to GVC position. I use the following method to calculate capital endowment index:

Kend = Capital stock / Population

Both the capital stock and population data are drawn from PWT 8.1 database.

4.5 Human Capital Index (HC)

To capture human capital development, I use variables drawn from PWT 8.1 dataset. The dataset provides Index of human capital per person, based on years of schooling (Barro & Lee 2012) and returns to education (Psacharopoulos 1994).

4.6 Trade Openness (Topen)

To capture trade openness, I use the proportion of total trade volume of goods and services. The data is drawn from the World Bank WDI database.

Descriptive statistics of all variables are shown in Table 1.

Table 1. Descriptive Statistics

Statistic	Measurement	N	Mean	St. Dev.	Min	Max
GVC Position	NA	6,274	-0.2	0.1	-0.7	0.2
IPR	NA	6,300	29.5	8.5	7.1	45.0
KEND	Mil US$ / person	6,300	75.1	43.4	3.2	233.8
HC	NA	6,300	2.9	0.4	1.7	3.6
Topen	%	6,300	85.0	52.9	20.3	349.8
PINT	NA	6,300	0.02	0.03	0.000	0.2

5. RESULTS

5.1 Baseline Estimation Results

Table 2 presents fixed effect estimates of baseline specification, equation (1). I focus on the coefficients of effective IPR protection, patent intensity and their interaction term. The coefficient of effective IPR protection is positive and statistically significant in all three estimations, showing that countries with higher IPR protection have higher GVC position. Patent intensity index has a negative impact on GVC position, while the coefficient of interaction term is positive. This indicates that industries with higher patent intensity need stronger IPR protection

Table 2. Main Estimates

	(1)	(2)	(3)
IPR	0.081*** (0.016)	0.051*** (0.017)	0.055*** (0.017)
PINT	-0.060*** (0.014)	-0.041*** (0.015)	-0.027* (0.014)
KEND		0.028** (0.011)	0.045*** (0.011)
HC		0.534*** (0.106)	0.539*** (0.103)
Topen			-0.106*** (0.005)
IPR*PINT	0.020*** (0.004)	0.014*** (0.004)	0.010** (0.004)
KEND*HC		-0.047*** (0.010)	-0.052*** (0.010)
Individual effect	Yes	Yes	Yes
Time effect	Yes	Yes	Yes
Observations	6,274	6,274	6,274
R2	0.084	0.089	0.146
Adjusted R2	0.075	0.080	0.131
F Statistic	43.057*** (df = 12; 5634)	36.731*** (df = 15; 5631)	60.142*** (df = 16; 5630)
Note:	**p*<0.1; ***p*<0.05; ****p*<0.01		

5.2 Dynamic Panel Estimates

Dynamic panel estimate results are presented in table 3. If lagged dependent variables also appear as explanatory variables, strict exogeneity of the regressors no longer holds. So I use instrumental variable (IV) approach to estimate equation (2), using second order lag of dependent variable as instrument. In Table 3, I also focus on the coefficients of effective IPR protection, patent intensity and their interaction term. In consistency with Table2, the coefficient of effective IPR protection is positive and statistically significant in all three estimations. The coefficient of patent intensity remains negative and statistically significant. The coefficient of the interaction term remains positive and statistically significant. Another notable point is that all adjusted R squares are above 2.6, much larger than those in Table 2.

Table 3. Dynamic Panel Estimates (IV)

	(1)	(2)	(3)
GVC Position lag	0.333***	0.344***	0.298***
	(0.028)	(0.028)	(0.029)
IPR	0.109***	0.047**	0.052**
	(0.020)	(0.022)	(0.022)
PINT	-0.092***	-0.054***	-0.050***
	(0.016)	(0.017)	(0.017)
KENT		0.072***	0.082***
		(0.013)	(0.012)
HC		0.741***	0.700***
		(0.112)	(0.111)
Topen			-0.090***
			(0.007)
IPR*PINT	0.029***	0.018***	0.016***
	(0.005)	(0.005)	(0.005)
KEND*HC		-0.068***	-0.073***
		(0.011)	(0.011)
Individual effect	Yes	Yes	Yes
Time effect	Yes	Yes	Yes
Observations	5,015	5,015	5,015
R2	0.300	0.309	0.326
Adjusted R2	0.262	0.269	0.284
F Statistic	163.386*** (df = 11; 4376)	134.704*** (df = 14; 4373)	137.118*** (df = 15; 4372)
Note:	**p*<0.1; ***p*<0.05; ****p*<0.01		

5.3 Endogenousity

When multinational corporations enter a host country, especially in IPR sensitive industry, they may require host country to provide stronger protection. The host country, on the other hand, may strengthen its IPR protection in order to attract more FDI in high technology industry. So IPR protection may be an endogenous variable in both equation (1) and (2).

The ideal way to prevent bias caused by endogenousity is IV regression. But as pointed out by relating literature, it's difficult to find good outside instruments for IPR protection. So I replace IPR protection in Table 3 with its lag to prevent two-way causality. Table4 presents the results. They are no difference with the results in Table 3.

5.4 Grouped Estimates

To test the robustness of the results, I divide the sample into two groups according to patent intensity of industries. Estimate results are shown in first two columns in Table 4. For both higher and lower patent intensity samples, the effect of IPR protection, patent intensity and their interaction term are similar to those in Table 2 to 4.

The last column in Table presents estimate result using random selected industry group sample. The coefficients are also similar to previous estimates.

6. CONCLUDING REMARKS

I investigated whether stronger IPR protection can improve a country's GVC position. Among previous literature, few provide empirical evidence that IPR protection help to improve a country's GVC position. Using a panel of up to 18 manufacturing industries in up to 35 economies between 2000 and 2009, this paper found evidence that stronger intellectual property protection is associated with higher global value chain position. The impact of stronger effective IPR protection is economically significant, and is stronger for higher patent intensity industries. The results are shown robust after IV dynamic panel regression, controlling for endogenousity and grouped estimation.

Table 4. Endogenousity

	(1)	(2)	(3)
GVC position lag	0.336***	0.347***	0.300***
	(0.028)	(0.028)	(0.029)
IPR lag	0.096***	0.045**	0.043**
	(0.016)	(0.017)	(0.017)
PINT lag	-0.082***	-0.053***	-0.035**
	(0.014)	(0.015)	(0.015)
KEND		0.080***	0.080***
		(0.013)	(0.013)
HC		0.779***	0.676***
		(0.117)	(0.115)
Topen			-0.090***
			(0.007)
IPR lag * PINT lag	0.025***	0.016***	0.011***
	(0.004)	(0.004)	(0.004)
KEND*HC		-0.073***	-0.072***
		(0.011)	(0.011)
Individual effect	Yes	Yes	Yes
Time effect	Yes	Yes	Yes
Observations	5,015	5,015	5,015
R2	0.300	0.309	0.324
Adjusted R2	0.261	0.269	0.283
F Statistic	163.578*** (df = 11; 4376)	135.071*** (df = 14; 4373)	136.433*** (df = 15; 4372)
Note:	*$p<0.1$; **$p<0.05$; ***$p<0.01$		

REFERENCES

- Awokuse, T. O. and Gu, W. G (2013). Does foreign intellectual property rights protection affect US exports and FDI? [J]. *Bulletin of Economic Research.*
- Bilir, K. (2011). Patent Laws, Product Lifecycle Lengths, and the Global Sourcing Decisions of U.S. Multinationals[J]. *SSRN Electronic Journal.*
- Branstetter, L., Fisman, R., Foley, C. F. (2005). Do stronger intellectual property rights increase international technology transfer? Empirical evidence from US firm-level data[R]. N*ational Bureau of Economic Research.*
- Branstetter, L. and Saggi, K. (2011). Intellectual Property Rights, Foreign Direct Investment and Industrial Development[J]. *The Economic Journal, 121(555),* 1161-1191.
- Branstetter, L., Fisman, R., Foley, C. F., et al. (2011). Does intellectual property rights reform spur industrial development?[J]. *Journal of International Economics, 83(1),* 27-36.
- Della Malva A. and Santarelli E. (2014). Intellectual property rights, distance to the frontier, and R&D: evidence from microdata [J]. *Eurasian Business Review,* 1-24.
- Ghosh, A. and Ishikawa, J. (2013). Trade Liberalization, Absorptive Capacity and the Protection of Intellectual Property Rights[J]. *UNSW Australian School of Business Research Paper.*
- Helpman E. (1993). Innovation, Imitation, and Intellectual Property Rights [J]. *Harvard Institute of Economic Research Working Papers, 61(6),* 1247-80.
- Hu, A. G. Z. and Png, I.P.L. (2013). Patent rights and economic growth: evidence from cross-country panels of manufacturing industries [J]. *Oxford Economic Papers, gpt011.*
- Ivus, O. (2010). Do stronger patent rights raise high-tech exports to the developing world? [J]. *Journal of International Economics, 81(1),* 38-47.
- Kalande and Christopher, M. (2002). Intellectual property, foreign direct investment and the least –developed countries. *Journal of World Intellectual Property, 5(1),* 11-128.
- Koopman, R., Powers, W., and Wang, Z., et al. (2010). Give credit where credit is due: Tracing value added in global production chains[R]. National Bureau of Economic Research.
- Lai, L.C. (1998). International intellectual property rights protection and the rate of product

Table 5. Grouped Estimates

	Higher PINT	Lower PINT	Random selected
GVC position lag	0.248***	0.334***	0.315***
	(0.050)	(0.034)	(0.045)
IPR lag	0.092***	0.085***	0.069**
	(0.035)	(0.023)	(0.029)
PINT lag	-0.062**	-0.094***	-0.067**
	(0.025)	(0.025)	(0.026)
KEND	0.093***	0.056***	0.086***
	(0.021)	(0.016)	(0.021)
HC	0.762***	0.411***	0.718***
	(0.191)	(0.138)	(0.188)
Topen	-0.083***	-0.093***	-0.085***
	(0.011)	(0.008)	(0.011)
IPR lag * PINT lag	0.020***	0.027***	0.019***
	(0.007)	(0.007)	(0.007)
KENT*HC	-0.089***	-0.042***	-0.076***
	(0.019)	(0.013)	(0.018)
Observations	2,511	2,504	2,512
R2	0.259	0.426	0.309
Adjusted R2	0.225	0.370	0.268
F Statistic	48.734*** (df = 15; 2181)	106.423*** (df = 15; 2176)	62.834*** (df = 15; 2182)
Adjusted R2	0.261	0.269	0.283
F Statistic	163.578*** (df = 11; 4376)	135.071*** (df = 14; 4373)	136.433*** (df = 15; 4372)
Note:	**p<0.1; **p<0.05; ***p<0.01*		

innovation [J]. *Journal of Development Economics, 55(1),* 133–153.

• Lippoldt, D. (2006). Can Stronger Intellectual Property Rights Boost Trade, Foreign Direct Investment And Licensing In Developing Countries? The Intellectual Property Debate: Perspectives from Law, Economics and Political Economy (ed. M. Pugatch), Edward Elgar Publishing, UK.

• Markusen, J. R. (2001). Contracts, intellectual property rights, and multinational investment in developing countries [J]. *Journal of International Economics, 53(1),* 189-204.

• Mathew, A. J. and Mukherjee, A. (2009). Intellectual property rights, southern innovation and foreign direct investment [J]. *Arijit Mukherjee, 31(2),* 128–137.

• Nabokin, T. (2014). Global Investment Decisions and Patent Protection: Evidence from German Multinationals[R]. Munich Discussion Paper.

• Noailly, J. and Ryfisch, D. (2015). Multinational firms and the internationalization of green R&D: A review of the evidence and policy implications [J]. *Energy Policy,* 83.

• Nunnenkamp, P. and Spatz, J. (2004). Intellectual Property Rights and Foreign Direct Investment: A Disaggregated Analysis [J]. *Open Access Publications from Kiel Institute for the World Economy, 140(3),* 394-415.

• Park, W., and Lippoldt, D. (2005). International Licensing and The Strengthening of Intellectual Property Rights in Developing Countries During The 1990s, OECD Economic Studies, 40.

• Pritchett, L. (1996). Measuring outward orientation in LDCs: Can it be done? [J]. *Journal of Development Economics, 49(2),* 307-335.

• Smith, P.J. (2001). How do foreign patent rights affect U.S. exports, alliate sales and licenses? *Journal of International Economics, 55,* 411–439.

• Tanaka, H. and Iwaisako, T. (2013). Intellectual Property Rights and Foreign Direct Investment: A Welfare Analysis [J]. *Discussion Papers in Economics & Business, 67(4),* 107–124.

• Wang, Z., Wei, S.J., and Zhu, K. (2013). Quantifying international production sharing at the bilateral and sector levels [R]. National Bureau of Economic Research.

• Yang, G. and Maskus, K.E. (2001) Intellectual property rights, licensing, and innovation in an endogenous product-cycle model[J]. *Journal of International Economics, 53(1),* 169–187

Can Low Risk Stocks Outperform High Risk Ones? Evidence from China Equity Market

Ruonan Ma
School of Finance, Nankai University, China

ABSTRACT

This paper tests the empirical relation between return and volatility on the Shanghai stock market. Herein, we find that the idiosyncratic risk anomaly is not significant in China, which can be attributed to the size and book-to-market value factors of the listing corporations. This is contrary to the typical evidence on the idiosyncratic volatility effect that low-risk stock portfolios may have a combination of low volatility and high returns. Furthermore, we investigate five market sections spanning diverse stock markets to confirm that low risk portfolios can hardly outperform high risk ones. Finally, we find that the low risk portfolios can stably get similar rewards to the high risk ones, especially in most fluctuating periods like the Global Financial Crisis.

Keywords: Volatility anomaly; Stock returns; Asset strategy; Factor model

1. INTRODUCTION

Treynor (1961), Sharp (1964), Lintner (1965) and Mossion (1966) propose the Capital Asset Pricing Model (CAPM), which postulates that the expected return on a stock is linearly proportional to its systematic risk, measured by beta. Yet, such Nobel-prize winning theory is surprisingly hard to find in empirical test results, which can date to Black (1972). On the U.S. stock market, a flat or even negative relation between risk and return is documented in the papers of Fama and French (1992), Black (1993), Falkenstein (1994), Haugen and Baker (1975, 1991, 1996). Further, numerous following studies indicate that low risk stocks can earn more returns than the CAPM predictions or even higher risk ones, which is called 'the idiosyncratic volatility effect' and implies that the equity investors overpay for risky stocks.

History of studying the volatility effect is given in Table 1. Haugen and Heins (1975) provide the evidence that the negative relation between beta and return is significant respectively on the U.S. stock and bond markets, which shows that the practical existence of risk premiums in the capital market. Haugen and Baker (1991), Jogannathan and Ma (2003) all confirm that the asset strategy of pursuing higher returns and lower risk with a minimum variance portfolio can defeat the capitalization-weighted benchmark in U.S equity market. Later, Carvalho et al. (2012) draw the similar conclusion using Sharp ratio, when examining substitutes for the capitalization-weighted strategies in the world market. After controlling the size, value and momentum factors, Blitz and Van Vliet (2007) research on the data from 1986 to 2006 and find the top (low risk)-minus-bottom (high risk) alpha spread reach 12% per annum, which they refer to as the 'volatility effect'. Adrian and Rosenberg (2008) find the prices of risk are negative, and, further, adding the short-run and long-run volatility components into the three-factor model is favorable to Fama-French 3-factor model.

Interesting studies in this regard are that one can find the flat or negative relation between risk and returns within nearly all capital markets. Blitz and Van Vliet (2007) discover that the European and Japanese equity markets share similar properties. And they find an uncommon low correlation between the volatility effect in the developed and emerging country. Using the ex post standard deviation of the daily stock return and monthly return across 23 developed markets, Ang et al. (2009) find that the average return spread between low and high idiosyncratic volatility may reach 1.3% per month. They also prove the effect in each G7 country and consider the trading frictions, private information in their test. Later, including at least 60 monthly data for each market, Blitz et al. (2012) find 11 emerging markets come across the volatility anomaly, and the effect appears to grow stronger over time. Furthermore, Haugen and Baker (2012) explore 21 developed and 12 emerging markets and prove that securities bearing more risk may yield negative rewards.

Further, the particular researches towards volatility anomaly on China stock market present various conclusions. Yang and Han (2009) find the idiosyncratic volatility generates a negative relation with cross-sectional returns. Xu (2010) documents both the idiosyncratic volatility and skewness are direct proportional to the corresponding returns and are negative to the expected returns. Meanwhile, Chen et al. (2009) repudiate that aggregate volatility

Table 1. References to the Volatility Effect

Author(Year)	Research Purpose	Data			Risk-return relationship
		Country	Frequency / Period	VOL-Measurement	
Black, Jensen, Scholes(1972)	Model Empirical Analysis	U.S.	M;1931-1965	beta	Positive
Fama,MacBeth (1973)	Model Empirical Analysis	U.S.	M;1926-1968	beta	Flat
Haugen, Heins (1975)	Empirical test	U.S.	M;1926-1971	beta; stdev.	Negative
Falkenstein (1994)	Empirical test	U.S.	M;1965-1992	variance	Negative
Haugen, Baker (1991,1996)	Model Empirical Analysis	U.S.	M;1979-1993	FF; variance	Negative
Rouwenhorst (1998)	Empirical test	U.S., Europe (12 in all)	M;1980-1995	beta	Negative
Ang, Hodrick, Xing, Zhang (2006)	Empirical test	U.S.	M;1963-2003	Daily VIX index change	Negative
Clarke, De Silva, Thorley (2006)	Empirical test	U.S.	M;1968-2005	Covariance matrix	Negative
Blitz, Van Vliet (2007)	Empirical test	U.S., Europe, Japan	W;1986-2006	stdev; beta	Negative
Bali, Cakici (2008)	Empirical test	U.S.	D/M;1958-2004	FF3 resid.,GARCH	Ambiguous
Adrian, Rosenber (2008)	Model Empirical Analysis	U.S.	D;1962-2005	Short/long-run	Ambiguous
Ang, Hodrick, Xing, Zhang (2009)	Model Empirical Analysis	23 Developed	D/M;1980-2003	FF3 resid.; lagged stdev.	Negative
Haugen-Baker (2012)	Empirical test	21 Developed,			(0.007)
12 Developing	M;1990-2011	Historical volatility	Negative		0.011***
Blitz , Pang , Van Vliet (2012)	Empirical test	11 Emerging markets	M;1988-2010	Grouping vol.,beta, MAD	Negative

and idiosyncratic risk are needed to be included in the pricing model after controlling the illiquidity.

As shown in Figure 1, where both the low and high risk portfolios from China stock market can bring higher returns than the medium-risk group, the relationship between risk and return is not fairly clear. Thus in this paper, we specifically focus on the volatility effect, considering the idiosyncratic risk anomaly on the Shanghai stock market, in order to further capture the actual risk-pricing principle in China.

Our main contribution to the existing literature is investigating the idiosyncratic volatility anomaly on the China equity market, specifically considering the industrial-level differences and fluctuating periods. Furthermore, we utilize the parsimoniously sorted portfolios and measurements, which can probably promise us more accurate results.

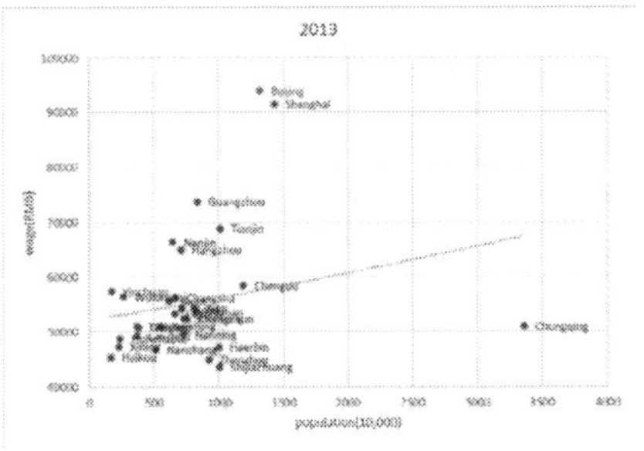

Figure 1. Average Annualized Return and Volatility for the Portfolios Sorted on Varied Risk (All A/B Shares in China, Jan.2012-Jan.2015)

The remainder of the paper is organized as follows. In the first section, we describe our data and methodology. Subsequently, we come to the primary focus of this paper, which is to empirically examine the volatility anomaly in the Shanghai stock market. We also control other effects (e.g. size, value and momentum),

Table 2. Summary Statistics of the Volatility Model Components (Shanghai stock market, Mar.1998 to Mar.2015)

	Mean	Median	Maximum	Minimum	Std. Dev.	Skewness	Kurtosis	Jarque-Bera
MKT	0.15%	0.45%	27.62%	-28.57%	0.08	-0.29	4.58	19.93
SMB	0.69%	0.41%	22.24%	-8.72%	0.04	1.27	9.24	319.72
HML	0.39%	1.16%	9.59%	-29.89%	0.05	-2.46	13.65	968.83
Rm	0.37%	0.68%	27.81%	-28.28%	0.08	-0.28	4.57	19.50

Notes: MKT, SMB, HML are respectively the excess return, the size and book-to-market value factors on the Shanghai stock market. Rm is the market portfolio return, substituted by the Shanghai composite index in our paper.

while specifically conducting the empirical tests with different frequencies and sub-sectors from diversified industries (market sections) as robustness tests. Finally, we end with our conclusions and implications.

2. DATA AND METHODOLOGY

In this section, we firstly describe our data, and then illustrate our methodology to explore the volatility anomaly empirically on China stock market.

2.1 Data

It is a fact that 'daily price limits', which started since December, 1996 in the China equity market, has a striking impact on the fluctuation of stock price and its volatility. And further considering data sufficiency and the historical stock market plumbing since June 2015, we set our in-sample period as March 1998 to March 2015.

Figure 2. Line Plot of Shanghai Composite Index (Shanghai stock market, Dec.1990 to Apr.2015)

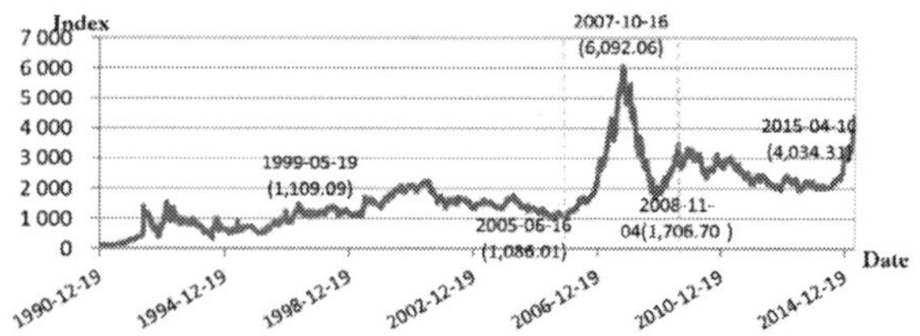

Herein, we remove the stocks of ST or *ST enterprises, which always come across a financial distress and probably have an abnormal excess return. The rest of the in-sample data are fully used. Our stock return data for the main research comprise monthly returns (ex. dividends) on 308 firms from the Shanghai stock market, which dominates the whole mainland equity markets. Firm-level data and individual stock returns are obtained from WIND and RESSET database. The returns used here are all in logarithm form. And we use log free-float market capitalization for measuring SMB (size, small-minus-big), inverse PB ratio for HML (book-to-market value, high-minus-low), and turnover ratio of shares for UMD (momentum). Summary statistics are given in the Table 2.

We compute excess returns using the one-year benchmark deposit rate, which is wildly used as the risk-free rate, designated as Rf, in China financial study and similar to the one-year bill yield of People's Bank of China (PBoC) in quantity. Also, we set the return rate of the Shanghai Composite Index as a proxy for the market return, designated as Rm. Thus we calculate 'Rm-Rf' as the risk premium on the China equity market.

2.2 Methodology

In this approach, we construct equally-weighted quintile portfolios by sorting the stocks on the past 36-month volatility of monthly returns at the end of each month, in accordance to the preceding empirical tests (Blitz & Van Vliet, 2007; Ang et al., 2009). And then we observe monthly average returns and Sharp ratio of the portfolios. In our analysis, transaction costs are ignored as one hypothesis.

As for statistically calculating volatility, researchers may resort to a beta, a standard deviation of realized returns or a mean absolute deviation (MAD). Goldstein and Taleb (2007) suggest using the MAD as a proper measure of the unsystematic risk, while Blitz and Van Vliet (2007, 2013) calculate the preceding volatility of a stock by taking the standard deviation of its monthly or weekly returns over the past three years. They further empirically prove that the simple historical volatility presents a stronger anomaly than the related beta. Baker et al. (2011) also draw the similar conclusion. Hence, we calculate the past volatility of a stock by standard deviation of its monthly returns over the preceding 36 months.

And such volatility will later serve as a main sorting benchmark in our research.

In addition, we provide the systematic risk factor beta and alpha estimated employing the CAPM, Fama-French three-factor (FF-3) and Carhart four-factor model regressions. Similar to the study of Rouwenhorst et al. (1999), Van der Hart et al. (2003), Blitz (2007) and Ang et al. (2009), the use of model regressions allows us to effectively control other anomaly effects (e.g. size, valuation and momentum). The one-factor alpha, Fama-French three-factor alpha and Carhart four-factor alpha are estimated using the following regressions:

$$R_i = \alpha_i + \beta_i MKT_t + \varepsilon_i \quad (1)$$
$$R_i = \alpha_i + \beta_i MKT_t + s_i SMB_t + h_i HML_t + \varepsilon_i \quad (2)$$
$$R_i = \alpha_i + \beta_i MKT_t + s_i SMB_t + h_i HML_t + u_i UMD_t + \varepsilon_i \quad (3)$$

Where $R_i = R_p - R_f$ denotes stock i's risk premium (also called as 'excess return') and $MKT_t = R_m - R_f$ is the risk premium for the market portfolio. SMB_t, HML_t, UMD_t are the returns on the size, value (book-to-market) and momentum factors. The systematic risk factor, designated as beta (β_i), can be obtained here, which presents the correlation between the individual portfolio and the market.

3. MAIN RESULTS

In the section, we utilize a sorting based methodology to test empirically our findings in the full-sample. Then we will investigate the robustness of our findings, considering the sub industrial market sections, sub periods and differed measuring frequencies.

3.1 Full-sample Results

The primary purpose is to find the relation of volatility and return in the Shanghai stock market during Mar.1998 to Mar.2015. The correlations of the variables are presented here.

As shown in the Table 3, we find the correlations between the variables are fairly low and thus our models have no striking multi-collinearity, except for SMB and HML. According to the Frisch Rule (1934), this can be proved later that their correlations are lower than the R-squared value and thus have no adverse effects. For the high correlation of SMB and HML in China, it indicates that small corporations always have a low PB ratio (the reciprocal of book-to-market, EPS/ book per share), which is accurately similar to what we can observe on China's main board.

Table 4 shows the monthly average returns and volatility of portfolios sorted on historical volatility, size, value and momentum factors. For each quintile portfolio, the top quintile Q1 contains most theoretical-favorable stocks with the lowest volatility, the smallest free float market value, the highest book-to-market ratio or the highest turnover ratio of shares, while vice versa. To avoid showing too many negative numbers, we calculate and present the spread of bottom-minus-top portfolio as Q5-Q1.

In the Panel A, we sort the portfolios on the historical volatility. Since the past volatility has a strong predicting power, the sorting method can be shown from the monotonically increasing standard deviation and the beta value, while the monthly average return fluctuates but has a rough increasing trend. This is against the idiosyncratic volatility effect. Considering the Sharp ratio, we could also find the bottom quintile Q5 is most favorable, with a ratio spread of 5.33% over the top quintile.

Further, the FF three-factor and Carhart four-factor alphas are quite different from one-factor alpha, indicating that the systematic size and book-to-market value exposures could explain the volatility effect in China. The four-factor alpha is slight lower than the three-factor one, suggesting the alphas might be attributed to the momentum effect. The Q5-Q1 return spreads of the portfolios sorted on the size, value and momentum in Panel B, C and D are all positive, which also explains the effects by themselves and indicates that we need not to report the regressed alpha values in the size, value and momentum sorting panels. Hence, we might confirm that the idiosyncratic

Table 3. Correlations of Variables
(Shanghai stock market, Mar.1998 to Mar.2015)

	MKT	SMB	HML	UMD	Std. Dev.	Skewness	Kurtosis	Jarque-Bera
MKT	1.00	-0.05	0.05	0.05	0.08	-0.29	4.58	19.93
SMB	-0.05	1.00	-0.74	0.39	0.04	1.27	9.24	319.72
HML	0.05	-0.74	1.00	-0.27	0.05	-2.46	13.65	968.83
UMD	0.05	0.39	-0.27	1.00	0.08	-0.28	4.57	19.50

Table 4. Portfolios Sorted on Volatility and Carhart Factors
(Shanghai stock market, Mar.1998 to Mar.2015)

	Q1	Q2	Q3	Q4	Q5	Q5-Q1
Panel A: Sorted on historical volatility						
Mean	0.43%	0.23%	0.28%	0.80%	1.21%	0.78%
Std. Dev.	7.29%	8.92%	9.79%	10.78%	12.11%	4.81%
Sharp ratio	2.84%	0.16%	0.64%	5.43%	8.17%	5.33%
beta	0.76***	0.92***	1.00***	1.09***	1.22***	0.46
t-value	22.67	21.78	20.78	20.23	19.79	-
1-factor alpha	-0.11%	-0.33%	-0.29%	0.21%	0.60%	0.71%
t-value	-0.41	-0.94	0.46	0.48	1.17	-
FF3-factor alpha	0.96%**	0.53%	-0.01%	-0.12%	0.00%	-0.96%
t-value	2.55	1.52	-0.03	-0.42	-0.01	-
Carhart4-factor alpha	0.95%**	0.53%	-0.01%	-0.12%	0.00%	-0.96%
t-value	2.53	1.52	-0.03	-0.42	-0.01	-
Panel B: Sorted on size						
Mean	0.62%	0.64%	0.85%	0.97%	1.11%	0.49%
beta	9.57%	9.74%	9.64%	9.46%	8.77%	-0.80%
Sharp ratio	4.21%	4.30%	6.56%	7.90%	10.18%	5.97%
Panel C: Sorted on value						
Mean	0.70%	0.66%	1.97%	0.74%	0.75%	0.05%
beta	9.04%	9.08%	18.31%	9.89%	9.90%	0.86%
Sharp ratio	5.38%	4.84%	9.57%	5.29%	5.40%	0.02%
Panel D: Sorted on momentum						
Mean	1.68%	1.83%	1.71%	1.75%	1.81%	0.13%
beta	9.61%	9.61%	9.77%	9.94%	9.04%	-0.58%
Sharp ratio	15.22%	16.81%	15.30%	15.39%	17.62%	2.40%

Notes: For each portfolio, we report the simple monthly mean of the log return, standard deviation of monthly return, Sharp ratio, CAPM beta, alpha and related t-statistics. Significance at the 1% level is denoted by ***, at the 5% level is denoted by **, and at the 10% level is denoted by*.

volatility anomaly may present contradicting results, which may ascribe to the size and value factors on the Shanghai stock market. That is, in China, the volatility effect might not perform as 'low risk high return'.
As additional empirical tests, we include equity data from other stock markets within China mainland, involving the Shenzhen stock market (Main and SME board respectively), and China's Chinext stock market. These tests all give out similar results.

3.2 Measuring the Sub-Market Sections

As the Table 5 shows, we further segment the whole China equity market so as to compare the effect in notably stable and fluctuant market sections. We choose the rather stable sections like materials, health service and transportation; also we observe the fairly active sections, such as IT and investment banking sections.

Since the stocks are sorted on volatility, we firstly find that the trend of standard deviation is in accordance with our quintile sorting rule. Further considering the sharp ratio, the high risk portfolio can bring rewards more efficiently in all industries observed here. It is apparent that the average return spreads of the top and the bottom portfolios show an apparent principle, that high risk can always generate high rewards.

3.3 Measuring the Sub-Period

To save space and present more evidently, we illustrate our findings using figures. As the Figure 2 in the Data part and Figure 3 shows, we separate the full-sample period into three sub-periods according to the historical fluctuation of Shanghai stock market.

And interestingly, we can still get the former conclusion as presented in the Figure 3, that is higher volatility portfolio (Q5) may always get a higher

Table 5. Portfolios Sorted on Volatility and Market Sections
(China equity market, Mar.1998 to Mar.2015)

	Q1	Q2	Q3	Q4	Q5	Q5-Q1
Panel A: Materials						
Mean	0.31%	0.26%	0.34%	1.01%	1.20%	0.89%
Std. Dev.	0.08	0.10	0.11	0.12	0.13	0.04
Sharp ratio	1.08%	0.47%	1.13%	6.78%	7.70%	6.62%
beta	5.12%	5.12%	4.80%	6.09%	3.16%	-1.96%
Panel B: Health Service						
Mean	0.87%	0.22%	0.75%	0.41%	1.45%	0.58%
Std. Dev.	0.08	0.09	0.10	0.11	0.13	0.06
Sharp ratio	8.55%	0.01%	5.22%	1.65%	9.25%	0.69%
beta	3.37%	8.11%	6.32%	1.35%	3.60%	0.23%
Panel C: Transportation						
Mean	0.17%	0.07%	0.30%	0.84%	0.85%	0.67%
Std. Dev.	0.07	0.08	0.09	0.11	0.13	0.06
Sharp ratio	-0.65%	-1.82%	0.85%	5.80%	4.73%	5.37%
beta	7.80%	6.07%	8.75%	-0.03%	8.00%	0.21%
Panel D: Information Technology						
Mean	0.04%	-0.25%	0.47%	0.75%	0.88%	0.83%
Std. Dev.	0.09	0.10	0.11	0.12	0.15	0.06
Sharp ratio	-1.99%	-4.69%	2.28%	4.61%	4.55%	6.54%
beta	7.89%	6.65%	8.04%	6.72%	24.21%	16.32%
Panel E: Investment Banking						
Mean	0.18%	0.18%	0.34%	0.71%	1.12%	0.94%
Std. Dev.	0.08	0.10	0.11	0.12	0.14	0.06
Sharp ratio	-0.50%	-0.43%	1.11%	4.09%	6.66%	7.16%
beta	5.23%	0.25%	2.47%	2.15%	6.46%	1.22%

Notes: We report the simple monthly mean of the log-return, standard deviation of monthly return, Sharp ratio, CAPM beta and related t-statistics.

reward. Moreover, for the dramatic fluctuating years, like 2007 to 2009, the return of Q5-Q1 (high-risk minus low-risk, bottom-minus-top) is stably positive but close to zero all the time. In relatively stable years like 2006 and 2012, one may witness the typical idiosyncratic volatility effect, performing as 'low risk high return', in the China stock market.

Figure 3. Spread of Monthly Average Return of Bottom-minus-Top Portfolio Sorted on Volatility
(Shanghai stock market, Mar.1998 to Mar.2015)

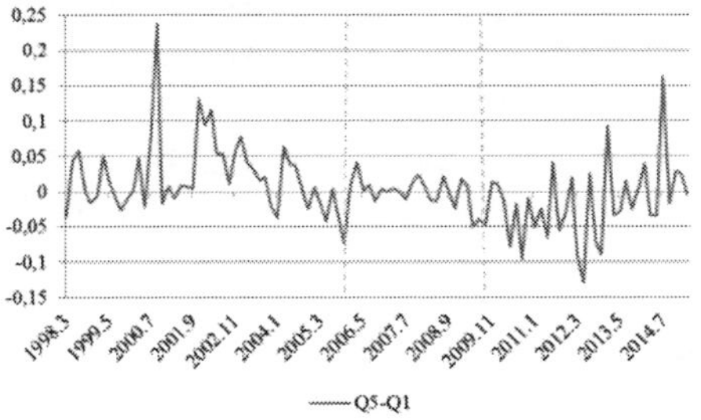

3.4 Differing the Data Frequency

We also diversify the frequency and length of our in-sample data. Using a trailing window of weekly return from 2014 to 2015, we can still obtain the same result. As Table 6 shows, after that we turn the weekly average return into the monthly measurement, we can still find the spread of Q5-Q1 (high-risk minus low-risk) reaches 1.24%, which means that investing in the high risk portfolio can bring a higher return due to weekly data. And a higher sharp ratio can also be achieved with high risk portfolios.

Herein, we find that our finding is robust to a different observing period and frequency for the return. Further computing can be carried out by substituting our log-return with the simple return rate, testing the effect on the observation of certain month, diversifying portfolio assets, or changing the equal-weighted

Table 6. Portfolios Sorted on Volatility of Weekly Returns
(Shanghai stock market, Mar.2014 to Mar.2015)

	Q1	Q2	Q3	Q4	Q5	Q5-Q1
Mean	0.57%	0.84%	1.09%	1.32%	1.81%	1.24%
Std. Dev.	1.90%	2.51%	2.76%	3.02%	3.16%	1.26%
Sharp ratio	-0.03	0.09	0.17	0.23	0.38	0.40
beta	0.50***	0.69***	0.76***	0.84***	0.85***	0.34
t-value	6.59	7.10	7.01	7.28	7.12	-
1-factor alpha	0.09	0.25	0.45	0.75**	1.28***	1.20
t-value	0.44	1.01	1.64	2.53	4.23	-

Notes: The statistics showed here are all turned into monthly computed measures. Significance at the 1% level is denoted by ***, at the 5% level is denoted by **, and at the 10% level is denoted by*.

portfolios into value-weighted and investor-mimicking portfolios.

4. CONCLUSIONS AND IMPLICATIONS

In China, regarding to our empirical test, the volatility anomaly hardly generates a significant effect. Hence, we can still witness that high risk portfolio always earns a rather high return. And in our market the asset strategy like 'Q5-Q1' (buy high-risk portfolios and sell low-risk ones) can always get a positive premium every month, while the average returns of portfolios sorted on volatility nearly have no steady trend. For the segments of the stocks based on the varied industry market sections or differed observing years, our conclusion still stands.

Moreover, in this paper, when utilizing the sorting methodology towards stocks during varied testing periods, we find that low risk portfolios can stably get a similar return to high risk ones, especially in most fluctuating periods, like during the Global Financial Crisis (from 2007 to 2009). This indicates that during the crisis period low risk portfolios will be more favorable and the flight-to-safety phenomenon actually happens. Moreover, during such years, researching on the relation of the idiosyncratic risk and monetary policy is promising to make sense in the China equity market.

To make our research conclusion of China more objective and compelling, the research forward is suggested to take policies, trading frictions, leverage, and information dissemination etc. into consideration. Additionally, methods involving market skewness and business cycle are well advised to be test, in order to discover more about the volatility anomaly.

REFERENCES

- Ang, A. Hodrick, R. J., Xing, Y. et al (2009). High idiosyncratic volatility and low returns: International and further US evidence. *Journal of Financial Economics, 91(1),* 1-23.
- Baker, M., Bradley, B., and Taliaferro, R. (2014). The Low-Risk Anomaly: A Decomposition into Micro and Macro Effects. *Financial Analysts Journal, 70(2),* 43-58.
- Bali, T. G. and Cakici, N. (2008). Idiosyncratic volatility and the cross section of expected returns. *Journal of Financial and Quantitative Analysis, 43(1),* 29.
- Blitz, D., Pang, J., and Van Vliet, P. (2013). The volatility effect in emerging markets. *Emerging Markets Review, 16,* 31-45.
- Blitz, D, Falkenstein, E.G., and Van Vliet, P. (2013). Explanations for the Volatility Effect: An Overview Based on the CAPM Assumptions. Available at SSRN 2270973.
- Blitz, D. and Van Vliet, P. (2007). The volatility effect: Lower risk without lower return. *Journal of Portfolio Management,* 102-113.
- Clarke, R.G., De Silva, H., and Thorley, S. (2006). Minimum-variance portfolios in the US equity market. *The Journal of Portfolio Management, 33(1),* 10-24.
- Fama, E.F. and French, K.R. (2012). Size, value, and momentum in international stock returns. *Journal of Financial Economics, 105(3),* 457-472.
- Falkenstein, E. G. (1994). Mutual funds, idiosyncratic variance, and asset returns. Northwestern University.
- Huang, W., Liu, Q., Rhee, S.G. et al. (2009). Return reversals, idiosyncratic risk, and expected returns. Review of Financial Study, hhp015.
- Haugen, R. A. and Baker, N. L. (1996). Commonality in the determinants of expected stock

returns. *Journal of Financial Economics, 41(3),* 401-439.

- Rouwenhorst, K. G. (1999). Local return factors and turnover in emerging stock markets. *The Journal of Finance, 54(4),* 1439-1464.
- Van Vliet, P., Blitz, D., and Van der Grient, B. (2011). Is the relation between volatility and expected stock returns positive, flat or negative? Flat or Negative.
- Xiaojun Xu (2010). Corporate idiosyncratic risk and the equity return based on the arbitrage in the Chinese stock markrt. *Economic Management, 12,* 127-136.
- Lili Liu et al. (2009). Empirical research on the idiosyncratic risk. *Statistics & Decision, 11,* 131-134.

The Diminution of Tariff of International Trade towards Inequality on Southeast Asia Countries (2007-2014)

Yahya Sudrajat
Intern Researcher at CSIS Indonesia (Centre for Strategic and International Studies)

ABSTRACT

In order to implement blueprint of ASEAN Economic Community (AEC), Southeast Asia countries have been gradually doing the diminution tariff of international trade since 2007. This effort has been implicating the decreasing of each Southeast Asia countries' GDP. Yet, this decreasing has also been followed by the augmentation of gini coefficients of each Southeast Asia countries during the same time periods. Therefore, hypothesis of this research is there is significant influence between the decreasing tariff of international trade towards the increasing of gini coefficient in the countries of Southeast Asia. Hence, this research had been concerned to seek an answer of how the significance of diminution of tariff of international trade towards the augmentation of Southeast Asia countries' gini coefficient.
Keywords: tariff, international trade, GDP, gini coefficient, Southeast Asia

1. INTRODUCTION

Before the end of this year (2015), Southeast Asia countries would start ASEAN Economic Community (AEC). The program has actually been started on 2007 by gradually decreasing tariff of international trade by South East Asia countries (Blueprint AEC 2007, p.1). By the end of 2015, there would be free trade for each South East Asia countries. Based on its blueprint (2007, p.18), AEC has been aimed to improve and strengthen economic capacity of South East Asia countries.

In addition, according to the data of World Bank, ASEAN's economic capacity has been strengthening since 2007. Based on GDP growth standpoint, in spite of the global financial crisis, GDP growth of this region was -0,64% in 2008. It means that economic capacity of this region was strong enough. Because in 2007, GDP growth of South East Asia was -15%, it means that aggregate of Southeast Asia's GDP growth was 43.6% at same year. In 2010, its GDP growth was 24%, and then in 2011 the GDP growth was 15% and 16% (databank.worldbank.org). According to those data, it could be concluded that based on GDP growth standpoint, Southeast Asia has economically been straightening.

Graph 1. Southeast Asia's GDP Growth (2008-2014)

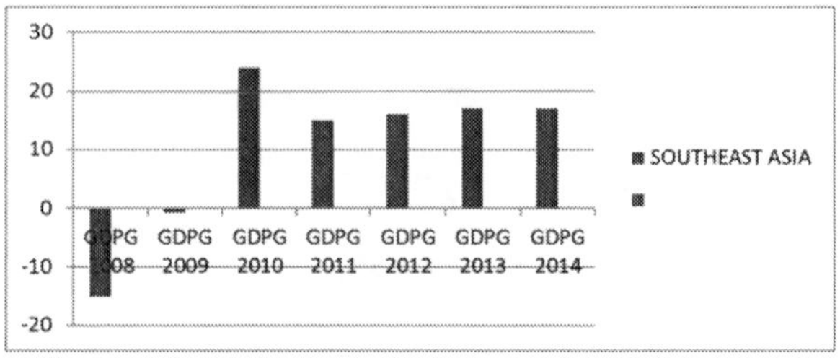

Note: The number is in billion USD
Source: databank.worldbank.org

Compared to the biggest emerging economic like China and India, the economic capacity of Southeast Asia has still been being competitive. For example, in 2010, China's GDP growth was 10,40%, and India's was 9,32%, and Southeast Asia's was still be competitive by 24% (databank.worldbank.org). The number of GDP growth of those countries has been continuing to 2014. Based on that fact, it could be concluded that economic capacity of Southeast Asia is big and competitive.

Graph 2. GDP Growth of Southeast Asia, India, and China (2008-2014)

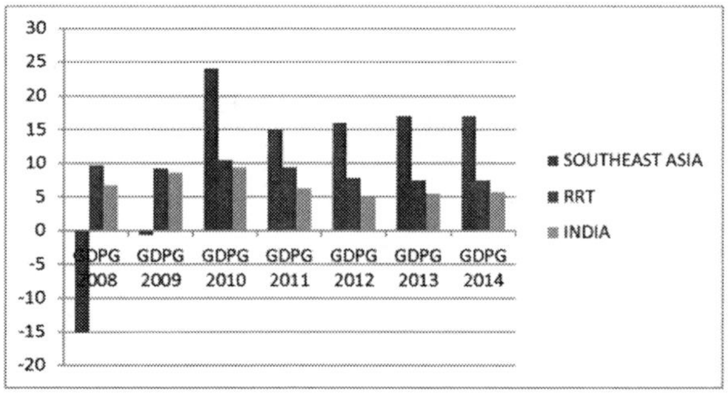

Note: The number is in percent (%)
Source: databank.worldbank.org

Based on another standpoint, according to GDP viewpoint, even though China's is always bigger, yet Southeast Asia's bigger than India's. For example, in 2007, GDP of Southeast Asia was 1,3 trillion USD, China's was 3,49 trillion USD, and India's was 1,2 trillion USD. In 2008, the pattern was the same, Southeast Asia's GDP was 1,5 trillion USD, China's was 4,5 trillion USD and India's was 1,29 trillion

USD. In the same year, GDP of China was bigger than Southeast Asia's, yet which was bigger than India's. The pattern has been being the same until 2014 (databank.worldbank.org).

Graph 3. GDP of Southeast Asia, India, and China (2007-2014)

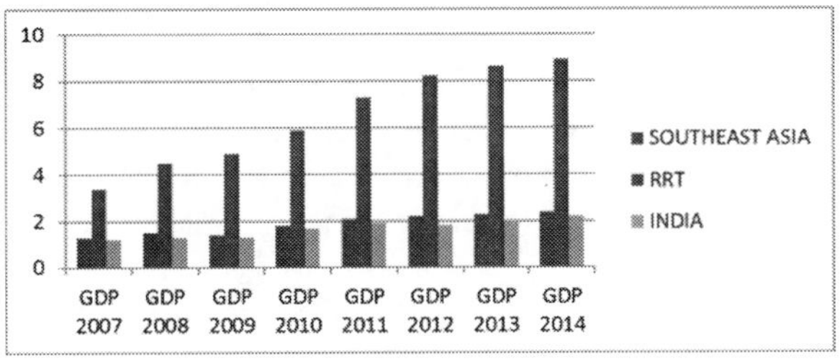

Note: Number is in billion USD
Source: databank.worldbank.org

Based on its total population, in 2007, Southeast Asia's total population was 573 million people. In 2008, its total population was 580 million. It means that the total population of Southeast Asia was significantly growing on that year. Furthermore, the pattern was still the same until 2014 (databank.worldbank.org).

According to PPP (purchasing power parity) viewpoint, the number of Southeast Asia's PPP was 11.463 USD in 2007. Compared to India's and China's PPP in the same year, PPP of Southeast Asia was bigger than China's, which was 3,620 USD and fewer than India's was 15.540 USD. Until 2014, its pattern was still the same, which means that Southeast Asia's was bigger than China's, yet India's was bigger than Southeast Asia's (databank.worldbank.org).

Graph 4. PPP of Southeast Asia, India, and China (2007-2012)

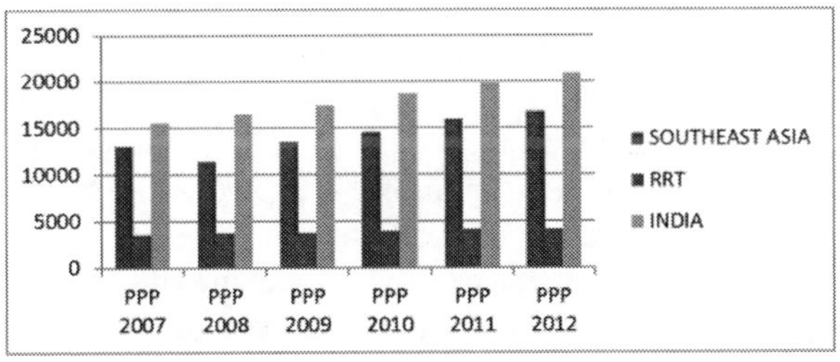

Note: The number is in USD
Source: www.databank.worldbank.org

Based on the explanation above, it could be concluded that the economic capacity of Southeast Asia has been being strengthening since the decreasing of tariff of international trade by Southeast Asia countries as an implementation of blueprint of ASEAN Economic Community. Yet, as strengthen as Southeast Asia economic capacity, gini coefficient of its countries has also been strengthen at the same period. For example, in 2007, Indonesia's economic growth was 6,3%. It was decreasing became 6% in 2008. So did its gini coefficient was decreasing at the same year from 3,4 in 2007, and become 3,0 in 2008. Yet while its economic growth was growing become 6,2% in 2009, its gini coefficient was also growing; became 3,8. The pattern had been continuous until 2014. These patterns had also been having in the others countries of Southeast Asia (databank.worldbank.org).

Therefore, the research is being focused to seek an answer of question of how significant of the decreasing of tariff of international trade towards inequality in Southeast Asia countries. Hence, to measure that significance, it is to be needed a regression analysis as an analysis tool. In addition, previous research and theoretical framework are also needed within this research to be a foundation of regression model, and theory for an analytical framework.

2. DISCUSSION

Within this research, the research of "The Impact of Trade Openness on Growth, Poverty, and Inequality on Vietnam" by CAO Xuan Dung (2004) became a foundation to build a model of regression. On his research, Xuan Dung (2004, pp. 8-12) has built model of regression as such.

Graph 5. Regression Model

Note: Openness: Tariff of International Trade
Growth: GDP Growth
Poverty: Poverty
Inequality: Gini Coefficient

There are three schemes of regression analysis on that research. First, the scheme of regression analysis, it was measuring the significance of tariff of international trade towards GDP Growth. Second, the scheme of regression analysis, it measure the significance of tariff of international trade towards poverty. And third, the scheme of regression analysis, it measure the significance of tariff of international trade towards gini coefficient (Dung 2004, pp.12-5). Hence, within this research, the third scheme is used, and the model of scheme of regression analysis it so measure the significance of tariff of international trade towards gini coefficient.

In addition, this research is aimed to measure the significance of the influence of the decreasing of tariff of international trade towards the inequality of Southeast Asia countries. Hence, model that would be used in order to measure significance of the influence both of tariff of international trade and inequality is below.

OPENESS ==>INEQUALITY

Note:
Openness = Tariff of International Trade
Inequality = Gini Coefficient

The Marxist Theory of Economic and Political Integration would be used to interpret and analyze the result of regression analysis of each Southeast Asia countries. The perspective of the theory is every free trade or economic integration would create the inequality of capital within the people of each countries. It caused by the competition of capitalist which resulting the exploitation by bigger capitalist toward fewer capitalist (Mandel 1970, p.88). At the beginning, the decreasing of tariff of international trade was influencing the increasing of economic growth of each Southeast Asia countries, yet the increasing of economic growth was influencing the decreasing of gini coefficient of each of Southeast Asia countries. To measure the coherency of the theory, we could see the result of regression analysis of each Southeast Asia countries beneath.

2.1 Indonesia

Based on the result of regression analysis of Indonesia, the significance of tariff of international trade towards gini coefficient is significant. The number of significance is 0,004, which means that the correlation between independent variable towards dependent variable is significant. In addition, it could be seen on the table beneath:

ANOVA[b]

Model		Sum of Squares	Df	Mean Square	F	Sig.
1	Regression	14.466	1	14.466	18.107	.004[a]
	Residual	4.794	6	.799		
	Total	19.260	7			

a. Predictors: (Constant), Y3
b. Dependent Variable: X3

The result of regression analysis above is coherence with the pattern of both of Indonesia's economic growth and gini coefficient data. For example, in 2009, Indonesia's economic growth was 4,6% and its gini coefficient was 3,8% in the same year. In the next year, that was 2010, the economic growth of Indonesia was 6,2% and its gini coefficient was 3,9% (databank.worldbank.org). Hence, it means that if there is the increasing of economic growth, it would be followed by the increasing of gini coefficient, and if there is the decreasing of economic growth, it would be followed by the deceasing of gini coefficient.

2.2 Laos

Regression analysis result of Laos is the number of regression 0,024. It means that tariff of international trade is significantly increasing gini coefficient in Laos. As like as Indonesia, the correlation of both variable is significant. Yet, the significance of it is not as significant as Indonesia. It could be seen beneath:

ANOVA[b]

Model		Sum of Squares	Df	Mean Square	F	Sig.
1	Regression	32.930	1	32.930	9.002	.024[a]
	Residual	21.950	6	3.658		
	Total	54.880	7			

a. Predictors: (Constant), Y6
b. Dependent Variable: X6

The pattern of economics growth and gini coefficient of Laos is coherence with the regression analysis above. For example, Laos' economic growth was 4,5% in 2008, and its gini coefficient was 3,2. In the next year, that was in 2009, its economic growth was 4,8%, and its gini coefficient was 3,4. It was meaning that both of the indicators were have the same pattern. The pattern was interesting in 2013 and 2014, its economic growth was 4,2% in 2013, and its gini coefficient was 4,2. Yet, in 2014, its economic growth was 3,5%, and its gini coefficient was 2,3. It was meaning that, if there was the decreasing of economic growth, it would be followed by the decreasing of gini coefficient (databank.worldbank.org).

2.3 Myanmar

The significance is also happened in Myanmar, the number of it is 0,006. It is mean that independent variable is significantly influencing dependent variable. Hence, the diminution of tariff of international trade is influencing the increasing gini coefficient significantly. As Laos and Indonesia,

Myanmar is also having the significance correlation. The result of regression analysis is beneath.

ANOVA[b]

Model		Sum of Squares	Df	Mean Square	F	Sig.
1	Regression	5.593	1	5.593	17.424	.006[a]
	Residual	1.926	6	.321		
	Total	7.519	7			

a. Predictors: (Constant), Y8

b. Dependent Variable: X8

Myanmar has the same pattern like the others. For example, in 2011, its economic growth was 3,3%, and its gini coefficient was 3,55. In the next year, that was 2012, its economic growth was 4%, and its gini coefficient was 3,56%. It was meaning that both of those indicators have the same pattern (databank.worldbank.org).

2.4 Thailand

The significance of the decreasing of tariff of international trade towards the increasing gini coefficient is 0,008. The number is significant, which mean that there is significant influence of tariff of international trade towards the increasing gini coefficient. It could be seen on the table below:

ANOVA[b]

Model		Sum of Squares	Df	Mean Square	F	Sig.
1	Regression	9.887	1	9.887	294.369	.008[a]
	Residual	.202	6	.034		
	Total	10.089	7			

a. Predictors: (Constant), Y7

b. Dependent Variable: X7

Thailand's economic growth was 7,5% in 2010, and its gini coefficient was 3,8 in the same year. Switch to 2011, its economic growth was 8,0%, and in the same year, its gini coefficient was 4,0 (databank.worldbank.org). It could be seen that both of indicators have the same pattern. If economic growth was increasing, then gini coefficient was increasing too. It is meaning that Thailand has also the same pattern like the others.

2.5 Philippines

The significance of the decreasing of tariff of international trade towards the increasing gini coefficient is 0,008. The number is significant, which mean that there is significant influence of tariff of international trade towards the increasing gini coefficient. It could be seen on the table below:

ANOVA[b]

Model		Sum of Squares	Df	Mean Square	F	Sig.
1	Regression	11.739	1	11.739	15.377	.008[a]
	Residual	4.580	6	.763		
	Total	16.319	7			

a. Predictors: (Constant), Y5

b. Dependent Variable: X5

For The Philippines, in 2008, its economic growth was 4,2%, and its gini coefficient was 3. Yet in the next year, that was 2009, its economic growth was 1,1% and its gini coefficient was 2,5. It was meaning that there was the same pattern of both of indicators on that period. In addition, in others periods, The Philippines economic growth was 3,7% in 2011, and in the same time, its gini coefficient was 4,0. In 2012, its economic growth was 6,7%, and gini coefficient of which was 4,2% (databank.worldbank.org). Hence, it is meaning that the pattern of both of economic growth and gini coefficient of The Philippines is coherence with the regression analysis above.

2.6 Malaysia

The significance of the decreasing of tariff of international trade towards the increasing gini coefficient is 0,043. The number is significant, which mean that there is significant influence of tariff of international trade towards the increasing gini coefficient. It could be seen on the table below:

ANOVA[b]

Model		Sum of Squares	Df	Mean Square	F	Sig.
1	Regression	7.172	1	7.172	4.318	.043[a]
	Residual	9.968	6	1.661		
	Total	17.140	7			

a. Predictors: (Constant), Y1

b. Dependent Variable: X1

In 2008, Malaysia's economic growth was 4,8%, and its gini coefficient was 3,6. Yet, in the next year, that was 2009, its economic growth was -1,5%, and its gini coefficient was 3,62% (databank.worldbank.org). It is interesting, because if there is the increasing of economic growth, there is also the decreasing of gini coefficient, and if there is the decreasing of economic growth, there is also the decreasing of gini coefficient. Hence, the pattern of it is coherence with the result of the regression analysis.

2.7 Vietnam

Regression analysis result of Vietnam is the number of regression 0,008. It means that tariff of international trade is significantly increasing gini coefficient in Vietnam. As like as Indonesia, the correlation of both variable is significant. Yet, the significance of it is not as significant as Indonesia. It could be seen below:

ANOVA[b]

Model		Sum of Squares	Df	Mean Square	F	Sig.
1	Regression	19.295	1	19.295	156.381	.008[a]
	Residual	.740	6	.123		
	Total	20.035	7			

a. Predictors: (Constant), Y9

b. Dependent Variable: X9

Vietnam economic growth was 5,4% in 2009, and its gini coefficient was 3,58 in the same year. In the next year, in 2010, its economic growth was 6,4%, and gini coefficient of the country was 6,4. Vietnam has the same pattern as the other countries, which have explained before that as the result of decreasing of tariff of international trade, the increasing of Vietnam's economic growth has been being followed by the increasing of gini coefficient. As Vietnam, Brunei Darussalam is also having the same pattern. For example, in 2010, its economic growth was 2,6% and its gini coefficient was 3,77. Switch to 2011, its economic growth was 3,4% and its gini coefficient was 3,79 (databank.worldbank.org).

2.8 Brunei Darussalam

Regression analysis result of Brunei Darussalam is the number of regression 0,043. It is mean that tariff of international trade is significantly increasing gini Coefficient in Brunei Darussalam. As like as Indonesia, the correlation of both variable is significant. It could be seen below:

ANOVA[b]

Model		Sum of Squares	Df	Mean Square	F	Sig.
1	Regression	7.172	1	7.172	4.318	.043[a]
	Residual	9.968	6	1.661		
	Total	17.140	7			

a. Predictors: (Constant), Y1

b. Dependent Variable: X1

Brunei was also having the same pattern like the others countries. For example, in 2010, its economic growth was 2,6%, and its gini coefficient was 3,77. Then, in 2011, its economic growth was 3,4%, and its gini coefficient was 3,79. It was also meaning that there is the same pattern in Brunei (databank.worldbank.org). Hence, it is coherence with the result of regression analysis.

2.9 Cambodia

Regression analysis result of Cambodia is the number of regression 0,0026. It is mean that tariff of international trade is significantly increasing gini coefficient in Cambodia. As like as Indonesia, the correlation of both variable is significant. Yet, the significance of it is also not as significant as Indonesia. It could be seen below:

ANOVA[b]

Model		Sum of Squares	Df	Mean Square	F	Sig.
1	Regression	70.778	1	70.778	8.699	0026[a]
	Residual	48.817	6	8.136		
	Total	119.595	7			

a. Predictors: (Constant), Y2

b. Dependent Variable: X2

For Cambodia, the pattern of economic growth and gini coefficient is coherence with the result of regression analysis. For example, its economic growth was 6,0% in 2010, and its gini coefficient was 3,0. And in the next year, in 2011, its economic growth was 7,1% and its gini coefficient was 3,8. The same pattern has also occurred in Myanmar, in 2009, its economic growth was 7,5%, and its gini coefficient was 3,51. In 2010, its economic growth was 8,0%, and its gini coefficient was 3,53 (databank.worldbank.org).

2.10 Singapore

Regression analysis result of Singapore is the number of regression 0,008. It is mean that tariff of international trade is significantly increasing gini coefficient in Singapore. The correlation of both variables is significant. It could be seen below:

ANOVA[b]

Model		Sum of Squares	Df	Mean Square	F	Sig.
1	Regression	9.887	1	9.887	294.369	.008[a]
	Residual	.202	6	.034		
	Total	10.089	7			

a. Predictors: (Constant), Y7

b. Dependent Variable: X7

Like the others Southeast Asia countries, the pattern of both of Singapore's economic growth and gini coefficient has the coherency with the result of regression analysis on the table above. For example, in 2007, Singapore's economic growth was 9,1% and its gini coefficient was 3,2. Yet in 2008, its economic growth was dropped, it was 1,8%, and its gini coefficient was also significantly decreasing was becoming 2,2. In addition, the pattern was interesting in 2010. On that year, its economic growth was significantly growing was 15,2%, and its gini coefficient was also significantly growing was 3,7 (databank.worldbank.org)

Based on the tables above, the result of regression analysis is significant for all countries. All of them are significant. It means that there is significant influence of the decreasing of tariff of international trade towards gini coefficient of those countries. Using Marxist theory of economic and political integration viewpoint, the increasing of gini coefficient means that the inequality of people within those countries is increasing too. It is coherence to the concept of that theory that free trade or economic integration would create the exploitation of "little capitalist" by "bigger capitalist", because there is capital competition among them (Mandel 1970, P88). Hence, the inequality within those countries is increasing.

Each of explanations above is showing that the decreasing of tariff of international trade is implicating the increasing the economic growth and it is implicating the increasing of gini coefficient. Based on the regression analysis above, the decreasing tariff of international trade is significantly influencing the increasing of gini coefficient in Southeast Asia countries. Hence, it could be concluded that the increasing of economic growth as the result the decreasing of tariff of international trade has been significantly implicating the inequality within the people in Southeast Asia countries.

3. CONCLUSION

Based on the regression analysis that has done, and the result is significant of all of those countries. Hence, the decreasing tariff of international trade, which implicating the increasing of economic growth, and it significantly influencing the increasing of gini coefficient of those Southeast Asia countries. In conclusion, based on this research, the decreasing of tariff of international trade is significantly influencing the increasing of gini coefficient within each of Southeast Asia countries.

REFERENCES

- www.databank.worldbank.org accessed at 11:43 PM 12/23/2015
- Mandel, E. (1970). Europe Versus America?: Contradictions of Imperialism. London: New Left Books.
- Dung, Xuan (2004). Online: The Impacts of Trade Openness on Growth, Poverty, and Inequality in Vietnam: Evidence from Cross-Province Analysis on www.afse.fr.

Session III: Global Dynamics

The Role of Industrial Cities in Diversifying the Economy The Case of Modon

Abdulrahman Abdulhameed Aljamie
Student, Al Yamamah University, Saudi Arabia

1. INTRODUCTION

Saudi Arabia is one of the largest oil exporters in the world, and its economy is highly depending on it. Knowing that oil prices are unsustainable, Saudi Arabia started to move toward more diversified sources of income by taking many initiatives. And one of those initiatives is Saudi Industrial Property Authority 'Modon'. Their main objectives are to build, regulate and promote industrial cities around the country and maximize their outcomes.

2. SAUDI ARABIA ECONOMIC OUTLOOK

Saudi Arabia is one of the largest and fastest growing economies with a nominal GDP of $750 billion in 2014, growing at 6.9% annually which puts it as the 19th largest economy in the world. Historically, oil was the main economic driver of the country, as the majority of the country's income came from oil exports. In the past ten years, oil prices were very high which fueled the country's growth. Despite that, the country knew the importance of diversity and the risk of dependency. Thus, it motivated all non-oil investments.

Saudi Arabia has a high and fast growing population with an average growth rate of 3% compared to the emerging market countries at 1.3% and the developed countries at 0.5%. The high population of Saudi Arabia helps the demand of products to be high and profitable. Also, Saudis who are younger than 25 make up more than half of the population, which will boost demand for products. Thus, more profitability to finish goods.

Figure1. Saudi GDP Growth (USD mn)

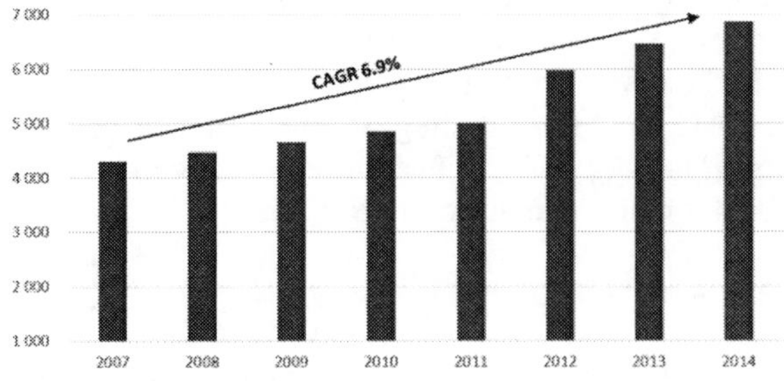

Source: Saudi Arabian Monitory Policy Annual report (2014)

Figure 2. Saudi Arabia Population Growth vs. EM and Developed

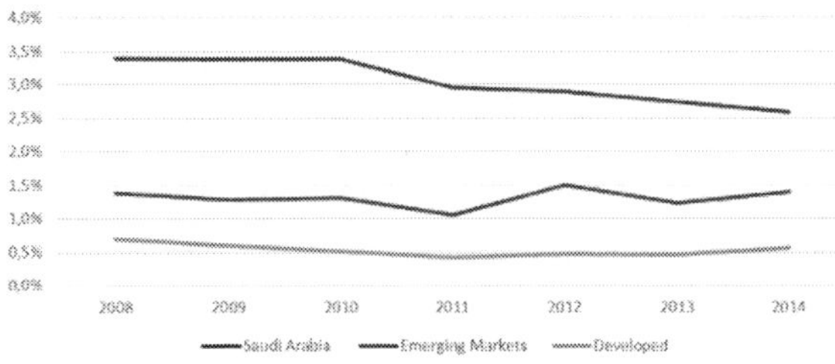

Source: Saudi Arabian Monitory Policy Annual report (2014), IMF data base (world economic outlook)

3. MODON

Modon is the industrial authority of Saudi Arabia. It was founded in 2001 and it is responsible for the development of industrial cities with integrated infrastructure and services. Modon's vision is 'to provide distinctive and unique industrial cities of the highest quality around the country'. Its main goal is to maximize the role of manufacturing in the country to support the growth of the Saudi economy. Toward that goal, Modon has established 34 industrial cities in different regions. These cities have around 6,900 factories with investment exceeding 150 billion dollars and more than 480,000 employees.

Since 2007, Modon had an aggressive plan to increase the factory count and workforce in their cities. Modon provided many incentives for investors, from cheap lands to access low cost financing. Modon leases lands to investors with an average of $0.50/sqm. Also, Modon provides pre-built factories for small and medium enterprises to decrease their starting costs. Furthermore, they facilitate financing through the Saudi Industrial Development Fund that covers 75% of the total factory cost with 20 years repayment period. Moreover, all factories are absolved from any customs charge for importing and exporting within the GCC region. Moreover, Modon will provide low cost fuel and energy to factories in their cities. Therefore, many investors received incentives to invest in the manufacturing sector, which increased it from 1,600 factories in 2007 to 6,900 in 2014 with a growth rate of 187 % annually.

Currently, there are 178 million square meters of developed lands with ready infrastructure for new factories. In addition, there are residential cities that are built next to the industrial cities with high quality houses, schools, hospitals and shopping centers to attract a talented workforce.

For factories to run, workforce is a vital element. However, a workforce cannot be employed unless they are well trained. The number of students in colleges of technical and vocational training has reached 113,900 in 2014. Prior to that, it was 70,000 in 2007. With an average of 23,000 graduates per year, factories will find well-trained employees to run them. These graduates will fulfill the increasing demand of workforce from factories.

The number of employees in the manufacturing sector more than doubled since 2007 coming from 420,000 employees to 935,000 employees. As a result, the contribution of oil on the GDP had an annual average of a 60% before the implementation. In 2014, the oil contribution decreased to 50%. The manufacturing sector increasingly contributed to the GDP until it reached 8% in 2014.

Modon plans to transfer a number of technologies to the Kingdom by 2020. To do so, it needs to start with downstream industries as these types of industries are simple and do not require much technical knowledge. Next, they will need advanced industries that already started with big investment in high tech assembly production lines. For example, Isuzu opened a truck assembly plant. This will help in the transferring of the truck building technology to people working in factories. The optimum goal of Modon is to have products fully engineered and made in Saudi Arabia. With these steps, the country will be able to do so in the future.

Figure 3. Students in Colleges of Technical and Vocational Training

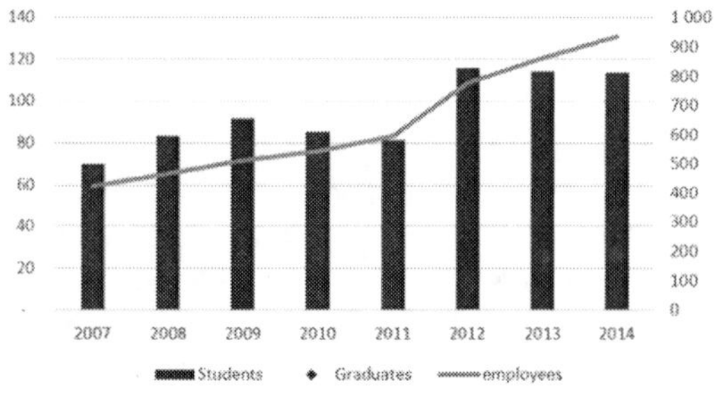

Source: Saudi Arabian Monitory Policy Annual report (2014)

Figure 4. Industrial Contribution to GDP

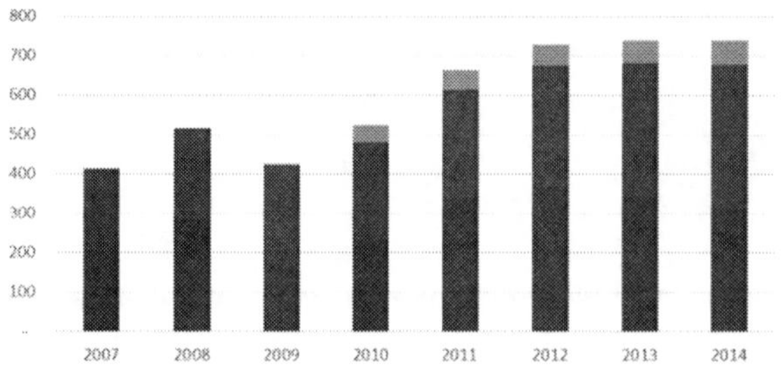

Source: Saudi Arabian Monitory Policy Annual report (2014), Modon Annual report (2014)

4. MODON LARGEST PROJECT WAAD ALSHAMAL (NORTHERN PROMISE)

Waad Alshamal project is one of largest projects in Saudi Arabia, and one of many projects that the government implemented to diversify sources of income. Waad Alshamal is one of the top five phosphate mining and processing projects in the world located in the north of Saudi Arabia. Part of the project is building a full industrial city with the infrastructure and residential area. The project has a total investment valued at $9.6 billion.

This site was found on the desert with no city near it. So Modon took a role in this large project to build the industrial city for the processing factories as well as the residential city for all workers and their families. In this city, Modon has developed the infrastructure and connected water and electricity to the city. Moreover, they connected Waad Alshamal with airports and harbors through railroads and highways.

5. CONCLUSION

Modon is a big organization that took responsibility to build an infrastructure for manufacturing in the country since 2001 to be a diversifying element in the oil depended economy. It started to act aggressively in 2007 by building cities and giving incentives to investors to invest in the emerging country of Saudi Arabia. As a result, it succeeded to put the manufacturing industry to 8% contribution of total GDP. After building 34 cities, it will continue providing support and facilitate financing to investors until it reaches their optimum goals.

REFERENCES

• Saudi Arabian Monitory Policy annual report (2014). Retrieved December 1, 2015, from http://

www.sama.gov.sa/ar-sa/EconomicReports/Pages/AnnualReport.aspx.

- International Monetary Fund: Economic Outlook. (October 2015). Retrieved December 1, 2015, from http://www.imf.org/external/pubs/ft/weo/2015/02/index.htm.
- Modon Annual report (2014). Retrieved December 1, 2015, from http://www.modon.gov.sa/ar/aboutmodon/achievements_authority/Pages/achievements_2014.aspx.
- Ministry of Economy and Planning: Saudi Economy in figures (2013). Retrieved December 5, 2015, from http://www.mep.gov.sa/themes/GoldenCarpet/index.jsp#1450793614109.
- Arab News (2014, February 13). Waad Al-shamal a future that won't rely on oil. Retrieved from http://www.arabnews.com/news/525351.

The Impact of Global Oil Price Decline on the Performance of Corporate Banking Sector in Saudi Arabia

Ahmed Ali Al-Yaagoub
Student, Al Yamamah University, Saudi Arabia
Corporate Relationship Manager, Gulf International Bank

1. INTRODUCTION

Money supply in the Kingdom of Saudi Arabia is witnessing a slower pace of growth affected by lower oil revenues. Many oil exporting countries rely on revenues from oil production to fund government spending. Falling oil prices will lead to a government budget deficit, and will require either higher taxes or government spending cuts. Saudi Arabia has built up substantial foreign currency reserves amounting to $ 700 Billion and can afford temporary falls in oil prices for some time. This is why Saudi Arabia has so far not responded by cutting oil output. Overall, the Saudi economy has continued to perform well despite the recent deterioration in oil prices, and all key real and financial macroeconomic indicators showed sustained improvement.

The Saudi financial system has gone through significant regulatory developments and continued to strengthen in recent years. The banking sector, which is the largest segment of the Saudi financial system, remained highly capitalized, profitable, and partially liquid. The sector remains connected to macroeconomic developments. So far, the recent drop in oil prices has not had an impact of the quality of the banking sector credit which is a result of the strong fiscal position, public debt sustainability, and government spending. However, such macroeconomic impacts call for close risk monitoring as history indicates some reactions by the financial system to oil market developments in the long run.

2. OBJECTIVE OF THE STUDY

This paper attempts to investigate the impact of global oil price decline on the performance of Corporate Banking sector in Saudi Arabia by examining the industrial and financial risks, as well as strategies and plans being implemented by Saudi banks on redefining of corporate clients segments and credit facility structures in order to recognize issues in a timely manner while adequate alternatives for remedial actions exist.

3. OVERVIEW

On Monday, the 28th of December, Saudi Arabia announced the budget for 2016 and reported the fiscal developments and budget implementation for 2015. The fiscal deficit for 2015 was announced significantly better than expected. The budget deficit came at SAR 367 billon (USD 98 billion), which is approximately 15% of GDP. The IMF expected the budget deficit to come at 21.6%, which was close to the market expectation. The biggest detractor in revenues appears to be oil revenues which are 23% lower than reported oil revenues in 2014. In 2015, oil revenues fell to SAR 444.5 billion and represented 73% of total revenues. On the other hand, higher expenditures were mainly a result of SAR 88 billion additional salaries (77% of budget vs. actual difference) paid in the year and SAR 20 Billion military spending (17% of budget vs. actual difference).

4. CORPORATE BANKING

Corporate Banking makes profits by taking small, short-term, relatively liquid deposits and transforming these into larger, longer maturity loans. This process of asset transformation generates net income for the commercial bank. Corporate Banking is critically important to the domestic and global economies as they make the loans that enable businesses to grow and hire people, contributing to expansion of the economy. The most immediate impact of current lower oil prices on Saudi banks is on the liability side followed by reduction of deposit inflows from large government and government-related entities that are causing system liquidity shortage of banks which leads to funding and liquidity pressures.

5. TARGET MARKET

Saudi Banks have recently redefined the parameters to select who to deal with, and why. Dealing with the right customer in the right industry is the cornerstone of a successful banking business. Experience has shown that through various Saudi banking crises, banks with a focused industry and customer selection strategy did better than other banks. Accordingly, defining the target market is the first step in creating well thought out risk process. Target market selection criteria for a bank are important across all banking segments, like Corporate and Retail Banking. Target market assessment is not only applied with existing customers, but also with prospect customers by measuring their company profile, products and services, and performance. In conclusion, target market is a set of guidelines to evaluate the following:

- Existing customers on annual basis to ensure there are no adverse developments within customers/ industry groups which would impact their existence with target market.
- Prospective customers at time of originating a new banking relationship to ensure they meet minimum credit standards.

5.1 Qualitative Factors of Choosing Corporate Clients

1	Ownership	Management performance, Ownership, Depth of Management, succession planning.
2	Industry competition and position	General Industry trend (growth, earning, suppliers, customers).
3	Reliability of forecasts and planning data	Historic Performance, future risk factors.
4	Earning development	Historic and future trend, compared to peer group within same industry, and influencing factors.
5	Financing / Cash Management	Exposure (debt volume, securities, earnings, source of repayment, future requirements).

6. DELAY IN PAYMENTS FOR SAUDI CONTRACTORS AND TIME DEPOSITS

Saudi Arabia is delaying payments to contractors assigned to work on government projects as the decline in oil prices pushes the country into a deficit for the first time since 2009. Companies working on infrastructure projects have been waiting for more than nine months for payments as the government seeks to preserve cash. Contractors had cash surplus in the past and were using these funds in deposits within Saudi banks, Saudi stock market, and other local investments. However, as they started witnessing delays in payments, they were forced to withdraw most of their deposits from Saudi banks. As a result, banks started having Assets and Liabilities mismatches, and were forced to pay higher interest rates to encourage existing and prospect clients to deposit their funds. Furthermore, this high offered rates on deposits led to increase in banks cost of funds, and as a result, banks started to increase the pricing of existing credit facilities for corporate clients. Payment delays have slowed the completion of projects under construction, including the USD 22 billion Riyadh metro project. Nevertheless, the Saudi government is still pushing contractors to complete the construction projects within the official contractual tenor. Therefore, contractors needed additional funding to complete their projects. Accordingly, Saudi banks have responded to this by financing the cash shortage with higher charges and credit facility pricing.

7. BUSINESS AND INDUSTRIAL RISKS WITHIN CORPORATE BANKING LENDING

Business Risks are inherent in a business or industry segment. They are external risks arising from the nature of the business, its operating environment and the economy. These risks are external to the company, but are affecting it. The corporate client of banks has limited control over the occurrence of these risks and their impact. Some businesses are inherently riskier than others, like construction and shipping companies. Cyclicality and volatility is associated with high business risk. Therefore, business risks are difficult to hedge or mitigate. Even if some business risks in a particular industry could be mitigated, they can never be eliminated. Financial Risks on the other hand are internal risks within companies and they are resulting from the financial and operating strategy followed, and the effectiveness of the company's operations. These are the risks banks are mostly using to assess the risks associated with companies by analyzing the financial spread sheet, and some ratios like current ratio, leverage ratios, liquidity ratio, and profitability ratio. Financial risks are usually easier to hedge and there is a certain degree of control over them. Therefore, most financial risks can be mitigated or fixed. In conclusion, these two types of risks should be inversely correlated, i.e. if there is a high degree of business risk, a low financial risk strategy should be chosen by a bank. Currently, Saudi banks are more focusing on analyzing business and industry risks in order to understand all the key risks and opportunities as well as determining the critical success factors for their Corporate and Retail lending portfolio, like growth prospects in relation to the Saudi economy, government regulations and policies effecting the industry, market share and position of corporations and individual clients.

8. FINANCIAL AND BUSINESS ANALYSIS

Financials analysis deals and focuses only on numbers of the annual audited financial, while business analysis deals with qualitative factors, business factors, customer's business strategy, customer's market share and positioning, and industry factors. However, most banks were focusing more on financial analysis rather than business analysis, which led to misunderstanding of current industry risks. However, banks need to know wither they are lending to clients within growing industry or not, and how the industry is growing in relation with clients activities. Furthermore, banks should know Saudi government spending strategy as banks should not deal with a client that only relies on government subsidies as he will not be able to perform once subsidies are cut.

9. HOW SAUDI BANKS PICK UP EARLY DEFAULT WARNING SIGNALS?

Banks can only pick up early warning signals by staying close to their clients with active contacts and frequent meetings, discussing business and credit issues, understanding customer's business strategy, understanding the industry and economic environment, and remain customer focused.

9.1 Early Warning Signals and Problem Recognition:

1- Excess capacity and supply like oil.
2- Government regulations and deregulations.
3- Technology changes.
4- Foreign competition and FX swings.
5- Sensitivity to economic cycles.
6- Change in consumer preference.
7- Poor position and market share within industry.
8- Financial results.
9- Government subsidies and spending cuts.
10- Sale of major assets.
11- Deteriorating trade relationships with suppliers and dealers.
12- High leverage ratios or low equity ratio.
13- Requests for extensions of loans maturities.
14- Negative operating cash flow and income.
15- Decrease in liquidity caused by short term obligations, which cannot be covered through cash flow from operations.
16- Larger overdrafts more than expected.
17- Negative legal records
18- Negative debt sourcing.

10. BANKS CLASSIFICATION OF CREDITS

Banks categorize the credit facility according to the risk of potential loss for the purpose of highlighting risks, evaluating the degree of risks involved, developing a collective action strategy, assessing the borrower's ability to rectify problem within a reasonable time, and considering of options available for improving the position of the creditor. Risk identification becomes much easier if banks have the best knowledge of their lending portfolio, and therefore they can easily assess the risk indicators. The objective of credit classification is to understand all possible risks and opportunities. This helps banks to know their target market and assess their current portfolio. Key risks in banking is all about understanding the business risks, and business risks can only be understood if bankers understand the industry where their clients are operating as risk analysis is a very important component of credit lending analysis.

10.1 Main Facility Structure Characteristics

1	Payment of Principle	Annuity, monthly, quarterly (Short Term debts to be paid through net operating cash flow)
2	Interest charges	fixed, floating, swapped
3	Tenor	Short, Medium and long Term. (avoiding any mismatch of assets against liabilities)
4	Collateral	Project proceeds, pledge of shares, personal and corporate guarantees.
5	Currency	domestic, FX

11. REMEDIAL MANAGEMENT STRATEGIES

The goal of remedial management is to get/recover the highest net present value of a loan. There are three strategies needed:

1- Recapitalization: banks try to get company owners to put additional equity, and other creditors also need to agree to similar strategy.
2- Rehabilitation: extended restructuring that could require a combination of new money, debt write off's, tenor stretching, renegotiation of terms, and new securities and collateral agreement.
3- Liquidation: aggressive strategy to recover the bank's money in accordance to the the following steps:

- Bank vs. the debtor/customer:

• Asses the problem (rate of deterioration)
• Identify all bank credit exposure.
• Asses the alternatives.
• Review documentation, security and collateral.
• Assessment of collateral liquidation value.

- Bank vs. other creditors/banks:

- Determine the position of other creditors.
- Determine the ability of existing client to arrange additional funds from other creditors to settle your credit exposure.
- Where does our bank stand versus other banks?

12. CONCLUSION

Based on this research, it is recommended for Saudi banks to ensure having the most proper industry and financial knowledge that will help them to diversify the corporate lending portfolios with clients that will survive, as well as identifying and recognizing all early default warning signals. Implementing the following three main strategies will enhance market penetration, gross selling, and healthy growth of lending portfolio:

Strategic Banking Goals	
1st Goal	1- Increase number of corporate clients
	2- Geographical expansion
	3- New products creation
	4- Market share growth
	5- clients and industries selection
	6- Profit targets increase
Market Segmentation and assessment of Corporate Clients	
2nd Goal	1- Volume of Growth
	2- Market Share
	3- Sector and client penetration
	3- Regional coverage
	5- Review of existing businesses
Performance Strategies for Banks	
3rd Goal	1- ROE/ROA/RAROC
	2- Monitoring level of provisioning.
	3- Loans to deposits ratio.
	4- Decrease number of weak clients

REFERENCES

- Fiedler, R.H. (2015). EUROMONEY, Corporate Client Relationship Management, 30.
- Saudi Arabian Monetary Agency (2015). Annual Report.
- IMF Country Report No. 15/286. ASSESSING THE IMPORTANCE OF OIL AND INTEREST RATE SPILLOVERS FOR SAUDI ARABIA
- Sarabjit Chhatwal (2016). Executive/ Managing Partner for BankT&D Training. *Problem Recognition & Remedial Management, 18,* 19, 48.

Spatial Wage Equation, Congest Effect and the Wage Gap

Zhanhan Yu
Master, Nankai University, P.R China

ABSTRACT

Traditionally, researchers use labor economics to explain the wage gap between cities in different size, which called urban wage premium. They think workers in bigger cities can receive higher wage because they are more capable and productive. Since wage must reflect such productive difference, therefore human capital premium leads to urban wage premium. However, this explanation may not be the whole story. Modern cities themselves could be the reason of wage gap between the cities. The agglomeration effect in metropolitan areas will make the firms more productive and thus afford higher wage and will allow the workers learning in the cities and thus make them more productive after living in the cities. In this paper, I am trying to use the panel data of 30 Chinese capital cities from 2004 to 2013 to demonstrate if there exist wage gap between Chinese cities and to explain what cause such wage gap.

Key words: wage gap; agglomeration economy; spatial wage equation

1. INTRODUCTION

It is a common phenomenon that wage in big cities is much higher than in smaller ones. Figure 1.shows the relationship between city size and its wage of 30 Chinese capital cities in 2013. We use population as a measure of the city size, and it shows clearly in the figure 1.that the higher level of city population, the higher average wage which workers in the city could receive[1]. Take Beijing and Shanghai as an example, they are two biggest metropolitan with over 1.2 million people and the wage in these two city are also the highest in China. As for Xining, Haikou and other small cities, the wage in there is much lower than in bigger cities.

Figure 1.The Relationship of Wage and City Size[2]

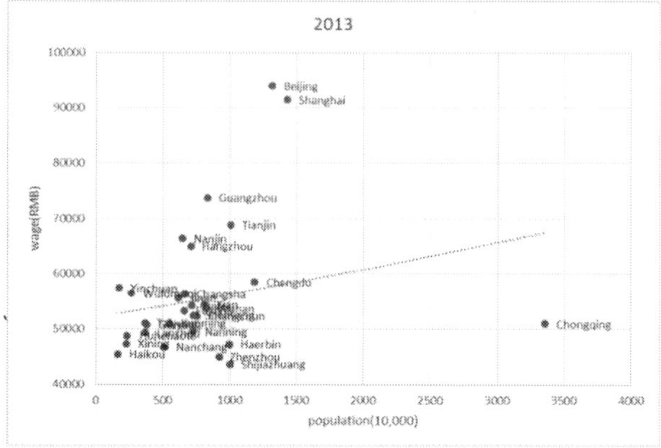

However, the positive correlation shows in figure 1.is not equal to a causal relationship. Is it because of the higher wage attract more workers to choose to find a job in bigger cities that leads to larger size of the bigger cities? Or it is because of firms in bigger cities could afford higher wage that make metropolitan areas have higher wage. Another possibility is that wage and city size interplay with each other and reinforce each other, which called self-reinforce effect in New Economic Geography theory. Traditionally, researchers use labor economics to explain the wage gap between cities in different size, which called urban wage premium. They think workers in bigger cities can receive higher wage because they have better education background and have more working experience, which means they are more capable and more productive. Since wage must reflect such productive difference, therefore human capital premium leads to urban wage premium.

Unfortunately, this explanation may not be the whole story. Modern cities themselves could be the reason of wage gap between the cities. When firms and people accumulate in the urban area, it will be much easier for workers to share the knowledge and creative ideas and to learn from each other. Also, for the firms in big cities, finding suitable workers is easier and transporting commodities could be much cheaper. The agglomeration effect in metropolitan areas will make the firms more productive and thus afford higher wage and will allow the workers learning in the cities and thus make them more productive after living in the cities. There are many research papers target at urban

[1] *Chongqing is an exception. As a direct-controlled municipality, Chongqing city includes some urban areas and a significant number of nonurban areas. The population data of Chongqing city provided by China City Statistical Yearbook represent both urban population and nonurban population. Therefore, the relationship of wage and urban population is biased in the sample of Chongqing city.*

[2] *Data come from China City Statistical Yearbook, 2014.*

agglomeration effect on urban wage gap. Glaeser and Mare (1994) first found there is urban wage premium existed between large cities and smaller ones. And they believed the higher product efficiency in large cities could be the main reason to explain the higher wage in there. But what is the mechanism of the agglomeration effect worked on cities? Helsley and Strange (1990) brought up a thick labor market theory (this theory also explained how agglomeration effect worked and such mechanism is called labor matching effect.), this theory then developed by Acemoglu (1997), Rotemberg and Saloner (2000), and Wheeler (2001). In their opinion, large cities have more population accumulated in the urban area and thus give more choice to the firms located in there. Also it will be much easier for the workers to find a satisfied job in big cities since more firms located in metropolitan areas. In other words, large cities have a thicker labor market which makes firms have more motivation to invest in new technology and on-the-job training. For workers, thicker labor market increase the possibility to find new jobs when lose their old ones, thus reduce their risk to make human capital investment. Both of the on-the-job training and human capital investment can raise city's product efficiency. Empirical study made by Yankow (2006), Wheeler (2006), Bleakley & Lin (2007), and Andersson, Burgess and Lane (2007) proved the thick labor market theory and showed that labor matching effect could be one explanation of urban wage gap between different sizes of cities.

As mentioned in Glaeser and Mare (1994), labor matching effect could only explain parts of urban wage premium. Urban wage premium will still exist after researchers controlled the labor matching effect, which means that there are other effects influence wage in cities. Another theory of the mechanism of agglomeration effect is called knowledge spillover and learning effect. Such theory could trace back to Marshall (1890), and Glaeser (1999) expanded and developed Marshall's idea to explain urban wage premium. What Glaeser concluded as one reason of the wage gap existed between various sizes of cities is the knowledge spillover and learning effect in cities. The geographic approaching makes people more likely to talk to each other and makes the information flow faster and easier. New ideas and new technologies expand quickly in cities which have higher population density because of people are approaching to each other. And in turn, when people communicate frequently to each other, the process of sharing novice ideas and common knowledge will promote the born of new idea and new technologies. People will learn from each other and in large city, the higher level of population provides more chance of learning. Those knowledge spillover externality and learning effect cause large cities and people working in there more productive. As we mentioned before, wage is a reflection of productivity, therefore, large cities always have higher wage. Researchers always use human capital externality to measure knowledge spillover and learning effect. Acemoglu and Angrist (2000), Ciccone and Peri (2006), Moretti (2004b), Gould (2007), Rosenthal and Strange (2008), and Combes, Duranton and Gobillon (2008), Graham and Melo (2009), Roca and Puga (2012) all use data from European or USA countries to estimate human capital externality effect on labor wages in urban area. But those empirical literatures have not reach a consensus that to what extent the human capital externality effect on labor wage. One main problem relies on this type of research is that it is based on partial equilibrium rather than general equilibrium. Wage is an endogenous variable in the whole economy system, thus partial equilibrium model is not enough to explain the wage gap among different sizes of cities. Another problem is the difficulty to find suitable and satisfied instrumental variables (IV). Such problem is a side-effect of partial equilibrium model: since wage is endogenous and many variables have effects on wage are also endogenous, researchers need to find IV as a proxy of those endogenous variables. Moreover, the micro-econometric study requires high quality micro data, which is hard to collect and find, especially in China. In 2005, Hanson provide a new method to use the spatial wage equation to explain the wage gap among cities. This method is based on the New Economic Geography (NEG) theory brought up by Krugman (1991) and its expansion model brought up by Helpman (1995). This model successfully avoids the problems mentioned before and provides a new perspective to explain urban wage premium.

In this paper, I am trying to use the panel data of 30 Chinese capital cities from 2004 to 2013 to demonstrate if there exist wage gap between Chinese cities and expand Hanson's model to explain wage gap among cities. The rest of paper will arrange as follows, next part will introduce the basic model of spatial wage equation and then I will use structural estimation method to do an empirical study. Results will illustrate in the third part and the final part is the conclusion.

2. MODEL AND METHOD

Since Krugman first brought up Core-Peripheral model (CP model) in 1991, Helpman (1995), Hanson (2005) and Brakman (1996) and Marrewijk (2004) expand and modified Krugman's model: Helpman used house market as a substitution of agricultural sector and Hanson use Helpman's model to build a spatial wage equation. On the other hand, Brakman and Marrewijk introduced congestion cost into the model as a new dispersion force of city. In this paper, I will try to use Hanson's method and combine congestion cost into classical CP model, and build a new kind of spatial wage equation. Using such spatial wage equation could allow us to analyze the reason why urban wage gap exists and factors that influence this urban wage gap. In the rest of this section, I will briefly illustrate my model.

Proposition one: consumers will buy two kinds of products - industrial product and house property. Therefore, we could write down utility function as follows:

$$U=H_r^{(1-\delta)} M_r^{\delta} \quad (1)$$

In equation (1), H represents house stock while M refers to the number of industrial products. $\delta\epsilon(0,1)$, represents how much income of consumers are they willing to spend on industrial products. Obviously, consumer's income could separate into two part Y=PH+IM, P and I are price of the product.

Proposition two: industrial firms an increasing returns to scale and in a monopolistic competition market. For the convenient of analyze, we assume that each firm only produce one product. The demand of industrial product can be written as:

$$M_r=(\sum_{(i=1)}^{N} c_i^{\rho})^{(1/\rho)} \qquad (2)$$

In equation (2), $\rho\epsilon(0,1)$ and refers to the product's elasticity of substitution. c_i and N represent differential industrial product and the number of firms. According to equation (1) and (2), we can calculate demand function as:

$$c_i=p_i^{(-\varepsilon)} I^{(\varepsilon-1)} \delta Y,\ i=1,2...N \qquad (3)$$
$$\varepsilon=1/(1-\rho)$$
$$M=\delta Y/I$$

$$\text{And } I \equiv [\sum_{1}^{N} p_i^{(1-\varepsilon)}]^{1/(1-\varepsilon)},\ \varepsilon>1.$$

Proposition three: assuming labor as the only factor used in production and according to Marrewijk (2004), we can write production function as follow:

$$l_{ir}=N_r^{(\varphi/(1-\varphi)}(\alpha+\beta x\_ir) \qquad (4)$$

N_r refers to industrial firms in location r and it could be seen as a proxy of city size. As people and firms accumulate in cities, the negative externality effect, such as congestion effect gradually appears. Traffic congestion, environment pollution and housing shortage is three main representation of urban congestion effect. In equation (4), we use external economies of scale coefficient φ to measure congestion cost. If $\varphi=0$, there is no external economies of scale; if $\varphi\epsilon(-1,0)$, there is positive external economies of scale, such as learning by doing spillover; and if $\varphi\epsilon(0,1)$, there exist negative external economies of scale, such as congestion effect we mentioned before. The profit function is:

$$\pi_r = p_r x_r - W_r N_r^{(\varphi/(1-\varphi)}(\alpha+\beta x_r)$$

W refers to industrial labor wage. Now we can use the profit maximum condition to calculate equilibrium price and equilibrium product.

$$x_r = (\alpha(\varepsilon-1))/\beta \qquad (5)$$
$$p_r=(\beta N_r^{(\varphi/(1-\varphi)} W_r)/\rho \qquad (6)$$

Proposition four: considering transport cost, we use ice-burg format which introduced by Samuelson. Transport cost can be seen as part of products which deliver from one place to the other. T_{rs} equals to the number of products which need to be transferred from location r to location s in order to make sure one unit of product reach location s. Thus:

$$T_{rs} = T^{(D_{rs})},$$
$$T_{rs} = T_{sr},\ T_{rr} = 0$$

Proposition five: assuming total labor is L, and industrial firms use γL labor and thus in location i, there are λ_i γL labor. From equation (4),

$$l_r=N_r^{\varphi/(1-\varphi)}\alpha\varepsilon, \text{ and } N_r = (\gamma L_r)/(N_r^{(\varphi/(1-\varphi)}\alpha\varepsilon)$$
$$\text{therefore, } N_r = (\gamma L_r^{1-\varphi})/\alpha\varepsilon.$$

Considering proposition one to five and short-term equilibrium (labor cannot transfer from one location to the others in a short time.), we can use these equations we have already get to calculate short-term equilibrium function of wage (W) and industrial product price (I):

$$W_s=\rho\beta^{-\rho}(\delta/((\varepsilon-1)\alpha))^{1/\varepsilon} (\gamma L/\alpha\varepsilon)^{-\varphi} \lambda_s^{-\varphi} [\sum_{s=1}^{R} Y_r T_{rs}^{1-\varepsilon} I_r^{\varepsilon-1}]^{1/\varepsilon} \qquad (7)$$

$$I_r=(\beta/\rho)(\gamma L/\alpha\varepsilon)^{(1-\varepsilon\varphi)/(1-\varepsilon)} [\sum_{r=1}^{R}\lambda_s^{1-\varepsilon\varphi} W_s^{1-\varepsilon} T_{rs}^{\varepsilon-1}]^{1/(1-\varepsilon)} \qquad (8)$$

Generally, we get the following simplified formula by normalization :

$$W_s = \lambda_s^{-\varphi} [\sum_{s=1}^{R} Y_r T_{rs}^{1-\varepsilon} I_r^{\varepsilon-1}]^{1/\varepsilon} \quad (9)$$

$$I_r = [\sum_{s=1}^{R} \lambda_s^{1-\varepsilon\varphi} W_s^{1-\varepsilon} T_{rs}^{1-\varepsilon}]^{1/(1-\varepsilon)} \quad (10)$$

Equation (9) is the Krugman's spatial wage equation which combine with congestion cost. This equation shows directly the relationship between wage and other regional variables such as income, product price and transport cost. However, the nonlinear format and the difficulty of collect regional product price make it not an ideal function for econometric estimation. What we need to do now is using Hanson's method to transfer function (9). First of all, define real wage:

$$w_s = W_s/P_s^{1-\delta} I_s^{\delta},\ w_r = W_r/P_r^{1-\delta} I_r^{\delta}$$

And in the long-term, the free movement of labor cause real wage becomes equal. Which means:

$$W_s/(P_s^{1-\delta} I_s^{\delta}) = W_r/(P_r^{1-\delta} I_r^{\delta}) \quad (11)$$

$$P_i H_i = (1-\delta)\, Y_i \quad (12)$$

Combining equation (9) (11) (12),

$$W_s = \lambda_s^{-\varphi} [\sum_{r=1}^{R} (Y_r^{(\varepsilon(\delta-1)+1)/\delta} T_{rs}^{1-\varepsilon} W_r^{(\varepsilon-1)/\delta} H_r^{(1-\delta)(\varepsilon-1)/\delta)}) / (Y_s^{((\varepsilon-1)(\delta-1))/\delta)} H_s^{((1-\delta)(\varepsilon-1))/\delta} W_s^{(\varepsilon-1)/\delta)})]^{1/\varepsilon} \quad (13)$$

Take the logarithm of (13),

$$log(W_s) = (-\varphi)log(\lambda_s) + \varepsilon^{-1} log(\sum_{r=1}^{R} Y_r^{(\varepsilon(\delta-1)+1)/\delta} T_{rs}^{1-\varepsilon} W_r^{(\varepsilon-1)/\delta} H_r^{((1-\delta)(\varepsilon-1))/\delta}) + \eta_s \quad (14)$$

$$\eta_s = -\varepsilon^{-1} log(\sum_{r=1}^{R} Y_s^{((\varepsilon-1)(\delta-1))/\delta} T_{rs}^{1-\varepsilon} W_s^{((\varepsilon-1))/\delta} H_s^{(1-\delta)(\varepsilon-1)/\delta}$$

Equation (14) is an estimable spatial wage equation, and if we define transport cost as $T_{rs} = e^{\tau d_{rs}}$, then we get estimation equation (15):

$$log(W_s) = (-\varphi)log(\lambda_s) + \varepsilon^{-1}\, log\, (\sum_{r=1}^{R} Y_r^{(\varepsilon(\delta-1)+1)/\delta}\, e^{\tau(1-\varepsilon)d_{rs}} W_r^{(\varepsilon-1)/\delta} H_r^{(1-\delta)(\varepsilon-1)/\delta}) + u_s \quad (15)$$

It is common in many researchers' studies to use reduced form equation instead of directly regressing spatial wage equation like equation (15). Reduced form equation could be simply written as:

$$\ln(W_s) = \beta_1 \ln(\lambda_s) + \beta_2 lnincome + \beta_3 lnhouseprise + \beta 4 harbor + \beta_5 edu + \ldots \quad (16)$$

Now we are going to use structural estimate method to execute non-linear regression in STATA. In order to eliminate other factors effect, we are going to add some control variables such as harbor (to control the influence of cities' location) and labor's educational background (to control human capital effect). We will compare the results of structural estimation with the OLS results to show which method can explain the wage gap between cities more precisely.

3. DATA AND RESULTS

First of all, income (*income*), wage (*wage*) and house price (*houseprice*) data come from China City Statistical Yearbook 2005-2014 and China Regional Economics Statistical Yearbook 2005-2014. Geography data such as cities' temperature (*temp*), precipitation (*rain*), average sunshine time (*light*) and humidity (*humid*), air quality (*airrate &PM10*) etc. come from China Statistical Yearbook 2005-2014. Next, we use the proportion of labors who received college or above education as a proxy of cities' human capital stock (*edu*), and we collect the data from China Statistical Yearbook 2005-2014. Finally, for the distance data, we calculate it by searching the shortest driving route in Google Map and use formula $d_{rr}=2/3\sqrt{(\pi area_r)}$ to calculate the distance within location r.

Reduced form equation in our paper could be written as:

$$\ln(W_s) = \beta_0 + \beta_1 lny + \beta_2 lnh + \beta_3 lnl + \beta_4 edu\ \beta_5 lnairrate + \beta_6 temp + \beta_7 rain + \beta_8 light + \beta_9 humid + u_s \quad (17)$$

And now we can use the OLS regression to estimate the panel data of Chinese 30 capital cities from 2004 to 2014. Figure 2 shows the results of the OLS regression. As we can see in column (1), income (in logs, lny) and house price (in logs, lnh) both have positive influence on wage, a surprise results compared to our analysis in part two. Educational background variable (edu) is significant and equals to 0.181, a result which accord with standard labor economic theory. Obviously, temperature (temp) and average sunshine time (light) influence urban wage level negatively, while precipitation (rain) and humidity (humid) is not significant, which means this two variables may not influence urban wage. Thus, in column (2) and column (3), we drop these two variables. According to column (1) and (2), air quality seems has no effect on city wage level, since variable lnairrate is not significant. However, column (1) and column (2) do not include variable PM10, an important measurement of air quality. When we add PM10 in column (3), it is clearly that estimated parameter of PM10 and lnairrate are -0.755 and -0.108, in 99.9% and 99% significance level respectively. The final results of OLS regression present in column (3). Income level and educational background of city labors are positive related with urban wage level. Estimated parameters of lny and edu are 0.486 and 0.140 in 99.9% significance level. The higher the income per capita, the higher wage in the city, and the more labors received college (or higher) education, the higher the city wage level. Average house price will also have a positive effect on urban wage level. The estimated parameter of lnh is 0.359 in 99.9% significance level. Air quality could also affect urban wage level: in our regression, the estimated parameter of lnairrate is -0.108 in 99% significance level. The better air quality of the city, the lower the cities' wage level. Notice that the parameter of lnl equals to -0.0529 in 99.9% significance level. Variable lnl represents city size, measured by the proportion of urban population in total population. City size and urban wage level are negative related with each other, proving the existence of congestion effect. These results suggest one serious drawback of OLS regress method in our model: it can only prove there exist a positive or negative relationship between house price (or air quality) and urban wage level, but cannot explain the internal mechanism lying in such relationship. Correlation does not necessary means causal relationship. In big cities, high level of house price and high level of wage always exist at the same time, we may need a more persuasive method to explain why such relationship exist.

Figure 2. Results of OLS Regression

	(1) lnwage	(2) lnwage	(3) lnwage
lny	0.491***	0.494***	0.486***
lnh	0.361*** (168.06)	0.358*** (81.59)	0.359*** (63.34)
lnl	-0.0558*** (-5.19)	-0.0547*** (-5.22)	-0.0529*** (-4.46)
edu	0.181*** (19.58)	0.182*** (6.28)	0.140** (3.02)
lnairrate	-0.0392 (-1.36)	-0.0382 (-1.36)	-0.108** (-3.01)
temp	-0.0861* (-2.14)	-0.0913 (-1.76)	-0.106 (-1.79)
rain	-0.00673 (-0.53)		
light	-0.101*** (-78.73)	-0.0887*** (-6.58)	-0.0937*** (-8.59)
humid	-0.0219 (-0.66)		
pm10			-0.755*** (-9.82)
_cons	3.194*** (8.86)	2.975*** (37.40)	3.526*** (47.23)
N	300	300	300
t statistics in parentheses			
* p<0.05, ** p<0.01, *** p<0.001			

In the next part of this section, we will use structural estimation method to conduct non-linear regression. Equation (18) is the spatial wage equation adding control variables. The structural parameters reported in Figure 3. is accord with the setting in second part of our paper.

$$\ln(W_s)=K_0 \ln lamda+\ln(\sum_{r=1}^{R} Y_r^{(\varepsilon(\delta-1)+1)/\delta} e^{\tau(1-\varepsilon)d_{rs}} W_r^{((\varepsilon-1))/\delta} H_r^{(1-\delta)(\varepsilon-1)/\delta})+K_1 edu+K_2 lnairrate+K_3 harbor+K_4 light+u_s \quad (18)$$

Figure 3. Nonlinear Regression Results

Variable	est3
KO	.30699228
sigma	1.6008984***
delta	.48477806***
tau	-.0206033***
K1	.969***
K2	.25853788***
K3	-.07608621
F4	-.23870133

In Figure 3 structural parameter sigma, delta and tau is significant in 99.9% significance level. K1 (=0.969***) is positive, which is much larger than OLS regression results. However, parameter K0 which measures congestion effect is not significant. Even though we find the evidence of positive externality in large cities, estimated parameter of air quality K2 still shows that congestion effect exist in large cities. The worse the air quality, the higher the wage level within that city. Bad environment situation reduce citizen's satisfaction degree towards cities they lived in and in order to compensate their psychological loss and attract them to work in the city, firms have to offer higher salary. Parameter K0 only measures cities' integrate externality level, therefore a positive K0 suggests that in large cities, positive externalities is dominant negative externality. In other words, agglomeration effect is much stronger than congestion effect in Chinese urban area. This explain why people chose to locate in large cities such as Beijing and Shanghai regardless of serious air pollution, traffic congestion and extremely high house price in those cities.

4. CONCLUSION

There indeed exist wage gaps between cities with different city size in China. According to our research in this paper, we find this gap is positive related to cities' income per capita and average house price. The higher proportion of workers who received college education in total population, the higher the wages. Coastal cities seem to have a higher wage, but we get similar results after we rule out the influence of these geographical features. Considering the relationship between urban wage and city size, the estimated parameter of city size is negative, suggesting the existence of congestion effect. For large cities, land shortage along with high land price increase house price. Unaffordable house price will make the city less attractive. In most of cases, higher house price in large urban area always be seen as a dispersion force, pulling citizens out of this city. To make sure the supply of labor in cities, wage in large cities must be higher enough to compensate the high house price. Similarly, as city size grows larger, things such as traffic congestion, air pollution and other environment problems as well as criminal issues could also make citizens feel less comfortable with living in the city. In this case, the wage difference between cities with different size is more like a compensation of negative externality of city congestion effect. Compensation of congestion effect is one explanation of urban wage gap. However, in our paper, we could not directly calculate agglomeration effect on firm's production efficiency through OLS regression, and therefore is not able to estimate to what extent the wage gap among cities influenced by agglomeration effect or more specifically, by firms' production efficiency. Even so, by using the structural estimation method, we could prove indirectly that the agglomeration effect and congestion effect interact with each other and both have an effect on the wage gap among cities. In our paper, we use the panel data of Chinese 30 capital cities from 2004-2014 to study how congestion effect and agglomeration effect working on wage gaps among cities with different city size. The conclusion of our study provides empirical evidence of Brakman's research and shed lights on the future study about urban wage premium.

REFERENCE

- Acemoglu, D. (1997). Training and innovation in an imperfect labour market. *The Review of Economic Studies, 64(3),* 445-464.
- Acemoglu, D. and Angrist, J. (2000). How large are the external returns to schooling? Evidence from compulsory schooling laws. NBER Macroecon Annul, 2000, 9-59.
- Baum-Snow, N. and Pavan, R. (2012). Understanding the city size wage gap. *The Review of economic studies, 79(1),* 88-127.
- Bleakley, C. H. and Lin, J. (2007). Thick-market effects and churning in the labor market: Evidence from US cities. Working paper, University of Chicago.
- Brakman, S., Garretsen, H., and Schramm, M. (2004). The spatial distribution of wages: estimating the Helpman & Hanson model for Germany. *Journal of regional science, 44(3),* 437-466.
- Breinlich, H. (2006). The spatial income structure in the European Union—what role for Economic Geography? *Journal of Economic Geography, 6(5),* 593-617.
- Ciccone, A. and Peri, G. (2006). Identifying human-capital externalities: Theory with applications. *The Review of Economic Studies, 73(2),* 381-412.
- De la Roca, J. and Puga, D. (2012). Learning by working in big cities. Working paper, CEMFI.
- Duranton, G. and Puga, D. (2004). Micro-foundations of urban agglomeration economies. *Handbook of regional and urban economics, 4,* 2063-2117.
- Garretsen, H., Brakman, S., and Schramm, M. (2000). The empirical relevance of the new economic geography: Testing for a spatial wage structure in Germany.
- Glaeser, E. L. (1999). Learning in cities. *Journal of urban Economics, 46(2),* 254-277.
- Glaeser, E. L. and Mare, D. C. (1994). Cities and skills (No. w4728). National Bureau of Economic Research.
- Hanson, G. H. (2005). Market potential, increasing returns and geographic concentration. *Journal of international economics, 67(1),* 1-24.
- Helpman, E. (1995). The Size of Regions.
- Krugman, P. (1991). Increasing Returns and Economic Geography. *The Journal of Political Economy, 99(3),* 483-499.
- Krugman, P. (1992). A dynamic spatial model (No. w4219). National Bureau of Economic Research.
- Marshall, A. (1890). Principles of economics, 8.
- Moretti, E. (2004b). Workers' education, spillovers, and productivity: evidence from plant-level production functions. *American Economic Review,* 656-690.
- Rosenthal, S.S. and Strange, W.C. (2008). The attenuation of human capital spillovers. *Journal of Urban Economics, 64(2),* 373-389.

• Rotemberg, J. J. and Saloner, G. (2000). Competition and human capital accumulation: a theory of interregional specialization and trade. *Regional Science and Urban Economics, 30(4)*, 373-404.

• Venables, A. J. (2011). Productivity in cities: self-selection and sorting. *Journal of Economic Geography, 11(2)*, 241-251.

• Wheeler, C. H. (2006). Cities and the growth of wages among young workers: Evidence from the NLSY. *Journal of Urban Economics, 60(2)*, 162-184.

• Yankow, J. J. (2006). Why do cities pay more? An empirical examination of some competing theories of the urban wage premium. *Journal of Urban Economics, 60(2)*, 139-161.

Law and Human Rights

Compulsory Voting: A Path towards More and More Equal Participation

Vicente Dinis Valentim
Political Science Master's Student, ISCTE - University Institute of Lisbon, Portugal

1. INTRODUCTION

Voter turnout levels have been steadily declining in the whole of the world's democracies, for the last decades. This finding is quite consensual among electoral behavior scholars. For example, Blais (2007) looks at 106 democracies throughout around 35 years, concluding that average voter turnout levels have begun to decline in the 1990's, and that the average level of such decline is 8 percentage points. This situation is even more problematic if we take into account that one of the major drivers of such a decline is the demobilization of young citizens, who vote at even lower rates than the rest of the population. Such fact has led some to predict a worsening of this problem in the future (Blais & Rubenson 2013).

This finding has led to an intensification of the debate around policies that may be able to reverse it. One of such policies is compulsory voting. As such, its implementation has been more frequently discussed recently, in countries such as United Kingdom, the United States, New Zealand or Jordan (Hill 2014).

In this paper, I will argue in favor of compulsory voting as a way of addressing low in voter turnout levels and, more specifically, of fighting the unequal political participation that such levels frequently imply. It should be noted, however, that the aim of the paper *is not* to discuss the arguments put forward for and against compulsory voting. Rather, it aims for the presentation of some recent empirical findings in the field of electoral behavior that, in my view, sustain the case for compulsory voting as a way of addressing political inequality. I also present some studies that suggest which penalties on nonvoters have a more beneficial effect, and should thus be imposed, if a country is to make voting compulsory in an effective manner.

2. THE SO-WHAT QUESTION: WHY SHOULD WE CARE ABOUT VOTER TURNOUT LEVELS?

In electoral democracies, the principle of 'one man, one vote' is one of most important manners in which to assure political equality, as it guarantees that the opinion of each individual is given equal weight, when deciding on the most important matters regarding the *polis*.

Does low voter turnout put such a principle under threat? Not necessarily. If the part of the citizens who vote were a representative sample of the whole population, we would not need to worry about turnout levels, because the issue opinions of voters would reflect those of nonvoters. To put it differently, we only need to worry about turnout levels as long as turnout is unequal, meaning that there are differences in relevant political and social characteristics of voters and nonvoters (Verba, Schlozman & Brady 1995). Unequal turnout is a problem, because it means that the concerns of different social groups are not addressed equally. When electoral participation is unequal, even if political actors are responsive to the views of voters, such views do not reflect those of the population *as a whole*. As Walter Burnham (1987, p. 99) as famously put it, *'if you don't vote, you don't count'*. This means that political actors may not feel the need to implement measures that address the concerns of social groups that permanently abstain from voting – if only because they are not aware of such concerns.

This problem of reduced responsiveness is even greater because it may create a vicious circle (Gallego 2014). Given that it leads members of low turnout groups to feel like their views do not matter as much as those of others, it can increase their dissatisfaction towards the functioning of democracy, and thus demobilize even the individuals of those groups that actually did vote (Lever 2009). This, in turn, gives those groups even less relative weight in each election, thus making the incentives of politicians to be responsive to their views even smaller. And cycle starts once more.

Therefore, it is crucial to understand whether or not turnout is unequal. That is what I shall turn to in the next section.

3. A DESCRIPITIVE ANALYSIS OF TURNOUT INEQUALITY

Most studies on electoral participation inequality have focused in the United States of America (USA). They are quite consensual as to the fact that, in that country, turnout is very unequal: people from groups with a lower socioeconomic status participate at consistently lower levels than those of groups with a higher socioeconomic status.

However, there is some controversy as to whether or not that finding can be replicated in other countries. In a robust recent study, Gallego (2014) shows that turnout inequality is neither a particularity of the United States case, as some authors have argued (e.g., Nie, Poweel & Prewitt 1969), nor ubiquitous, as others have defended (e.g., Fowler 2011). Taking formal education level as an indicator of socioeconomic status, the author rather shows how turnout inequality varies widely across the world, as Figure 1 shows us.

Figure 1. Predicted Probability to Vote by Education, in 85 Elections

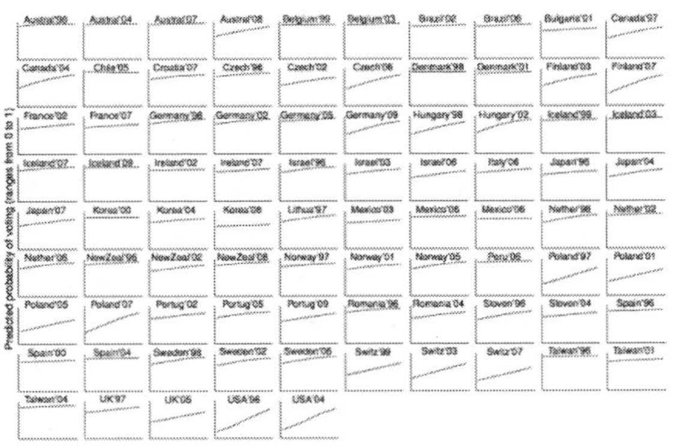

Source: Gallego, 2014

According to these results, the author divides countries included in her sample into four categories: one in which the association between education and voter participation is strong – USA, Poland, Canada, Finland, Hungary and Switzerland; one in which such association is medium – Austria, Israel, Japan, Lithuania, Portugal, Romania, Slovenia, Sweden, Portugal, Romania and the UK; one in which the association between education and turnout is weak or even nonexistent – Australia, Belgium, Brazil, Bulgaria, Chile, Denmark, Iceland, Korea, Peru, Spain and Taiwan; and one in which the size of such association varies widely from election to election, thus making its actual dimension hard to classify – Czech Republic, Germany, Netherlands, New Zealand and Norway. As we can see, even though there are a number of countries in which the level of inequality in political participation is not particularly worrying, most countries do in fact present considerable levels of such inequality. Furthermore, one should take into account that a number of countries in which participation inequality is not particularly worrying already have compulsory voting: Australia, Belgium, Brazil and Peru. Also, Chile had compulsory voting between 1925 and 2012. And, as we know, voting is path-dependent, meaning that people create a habit to vote or not to vote, and such habit takes some time before it is reversed (Franklin 2004). Therefore, one could hypothesize that the low levels of turnout inequality found in Chile are still the effect of the long period of compulsory voting it experienced. This being true, what this analysis shows us is that half the countries with equal or near-equal turnout are countries in which voting is already compulsory, or in which voting was compulsory for a long time and just recently became voluntary. Conversely, it leaves us with a small number of voluntary-voting countries in which turnout inequality is small or inexistent.

4. HOW TO MAKE TURNOUT MORE EQUAL: CONTEXTUAL CHARACTERISTICS WITH HOMOGENEOUS AND HETEROGENEOUS EFFECTS

To understand the ways in which one could aim for more equal electoral participation, in countries where such participation is unequal, it is useful to rely on the distinction, also made by Gallego (2014), between contextual factors that affect participation in heterogeneous and homogeneous ways (see also Verba, Nie & Kim 1978). The former disproportionally affect citizens of a specific socioeconomic group, whereas the latter affect citizens of all social groups in the same manner. For example, rainy election days or long waiting lines are likely to have a homogeneous negative effect on turnout, because one can expect individuals of different socioeconomic status to be equally demobilized by it. On the other hand, factors such as difficult registration procedures are likely to demobilize underprivileged citizens ate higher rates, as these have a smaller number of resources that enable them to cope with the challenges posed by such procedure.

This distinction has important implications for policy prescriptions, if one is to aim towards more equal participation. Both heterogeneous and homogeneous effects may be able to push towards that goal, even though they can do so in different ways. Contextual characteristics with heterogeneous effects can equalize participation if they disproportionately mobilize citizens from low socioeconomic status groups. On the other hand, contextual characteristics with homogeneous effects can equalize participation only if they bring the overall participation levels close to its maximum, because, when participation gets very close to 100%, the difference between the participation of different socioeconomic groups necessarily starts to shrink (Lijphart 1997). As the electoral participation of higher socioeconomic status groups gets close to its maximum, further increases in overall turnout are made at the expense of lower socioeconomic status individuals, thus making participation *more equal.* If electoral participation was actually 100%, there could be no difference between the participation of individuals of different groups, because *everyone* voted. However, compulsory voting is the only institution that can, on its own, have a homogeneous effect of such magnitude.

5. COMPULSORY VOTING

5.1 Why Compulsory Voting?

Whereas one can find several examples of contextual characteristics that can equalize participation through heterogeneous effects – such as vote-facilitating rules, for example, compulsory voting is the only institution able to have, on its own, a homogeneous effect that is strong enough as to make participation more equal. It is true that a great number of authors would call for other measures to be applied *before*, or even *instead* of compulsory voting. However, it is also true that the actual power of each of those measures to increase and equalize participation *on its own* is small. Also, those measures take a long time before their effects are actually felt. And even enormous investments in areas such as education would lead to relatively small increases in turnout levels (Franklin 2004). Therefore, compulsory voting has the great advantage of being the only policy able to, on its own, not only increase the overall levels of turnout, but also make it more equal. Needless to say, its financial costs on the state are small. Furthermore, compulsory voting can produce more and more equal participation even without the need to reach a consensus on the factors that make it both low and unequal (Hill, 2014). This is particularly important if we take into account that there is intense debate among political behavior scholars concerning the contextual factors that affect voter turnout, and that these scholars find it very difficult to disentangle such factors from each other, as most of them are highly correlated (for a good overview of this debate, see Blais 2006).

5.2 How and Where Should Compulsory Voting be Applied?

Naturally, compulsory voting does not make sense everywhere. It makes sense in countries in which turnout is unequal, and its pertinence and benefits are larger in countries with higher levels of such inequality. However, as we have seen, countries in which participation inequality is small or inexistent are the exception rather than the rule.

Also, being in favor of compulsory voting does not necessarily mean that we need to leave aside all other measures that bring about more equal turnout. In fact, I would argue that such measures are in fact a necessary complement for compulsory voting. The application of compulsory voting on its own is not able to magically fix all problems of political inequality. Naturally, there are some risks associated with its implementation. Authors that oppose to compulsory voting have made some important points regarding those risks. However, compulsory voting can be a huge step forward in terms of enhancing levels of political equality. And, by implementing it alongside other measures that fight political inequality, we can actually avoid its risks.

Let me give you an example. There is an ongoing debate between political behavior scholars concerning the relationship between compulsory voting and political knowledge and interest for politics. Some authors (e.g., Hoffman e Graham 2006) argue that, when people are compelled to vote, they will try to do so with as much information as possible, thus increasing the levels of political knowledge and interest for politics of the population. Others argue that, if people are not *a priori* interested in politics, compelling them to vote will do nothing more than increasing their alienation towards it (Ballinger 2006). According to this view, compulsory voting can thus create what, in Social Psychology, is called a reactance phenomenon: the situation in which a person feels that a certain liberty is being taken away from him or her, thus reacting by intensifying the behavior associated with the liberty he or she perceives as being under threat (Brehm 1995). In this case, when a person is compelled to vote and does not want to do so, he or she reacts by dragging further away from politics. This example illustrates my point of how compulsory voting should be implemented alongside other policies aiming towards an increase in people's interest for politics. If, together with making voting compulsory, we implement measures capable of engaging people into the public affairs of their countries, as well as measures that make the act of voting easier, we have a better chance of fulfilling its beneficial potential, increasing people's engagement in the *polis*, while avoiding its risks – namely, that of pushing people even further away from public issues. Some examples of those measures include civic education programs, vote-facilitating rules, etc. My aim here is not to discuss those measures (but, for a good discussion, see, for example, Hofflman & Graham 2006, pp. 111-3). My point here is, simply, that compulsory voting does not need to be applied *instead* of other measures capable of bringing about higher levels of turnout and political engagement on the behalf of citizens. On the contrary: its chances of success will actually increase if it is applied *alongside* such measures.

5.3 Sanctions on Nonvoters

It should be noted, however, that compulsory voting does not brings turnout levels close to 100% in all the countries in which it is applied. For example, voter turnout in 2015 Greece parliamentary election was no higher than 63.6% (IDEA 2016). This finding suggests that the effectiveness of compulsory voting in bringing about near-universal turnout rates depends on the conditions under which it is implemented. Central among those is the kind of sanctions imposed on nonvoters. Therefore, in this last section I shall draw upon those penalties. My goals here are twofold. First of all, I aim to present a typology of the sanctions that can be imposed on people who refrain from voting, as well as to provide an overall perspective of the ones that are more frequently used. After that, I will refer to the results of some empirical studies suggesting which of those sanctions are more effective in pushing turnout levels towards near-universal rates. This is very important, because I am arguing for compulsory voting as a way of reducing inequality in electoral participation. And, as we have seen, the way in which compulsory voting can accomplish that goal is through such near-universal rates. Therefore, when compulsory voting fails to do so, it fails to go through with its most important objective. That is why it is crucial to look at the conditions under which this policy has greater chances of enhancing electoral participation.

Gratschew (2004) draws a distinction four kinds of penalties that can be imposed on people who fail to vote, under compulsory voting. These are not mutually exclusive, and combinations of several kinds can, in fact, be found in several countries. This being said, the first kind referred by the author is the obligation of the person that refrains from voting to present a valid explanation of why he or she did so. Naturally, this penalty can be more or less strict, depending on the amount of explanations that are considered to be valid. The second kind of penalty is the obligation of the people who do not vote to pay a fine. Again, this penalty can assume several levels of harshness, depending on the value of the fine. The third type of sanction corresponds to the possibility of incarceration of the abstentionist. And the last one corresponds to the withdrawal of some of his or her civil rights.

This author presents also a list of countries in which voting is compulsory, or in which it was so in the past. If one takes into account the nations in which voting is currently compulsory, as well as those in which it was compulsory in the past but no longer is, and those in which the penalties applied on abstentionists have been changed, we come up with a total of 33 cases. Among these, 11 demanded an explanation on the behalf of the individuals who failed to vote, 19 applied fines on them, 3 considered the possibility of their incarceration, and 3 admitted the withdrawal of some of his or her civil rights. There are still 3 cases in which the penalties applied on abstentionists cannot be fitted into either of these categories, and 9 in which there is no sanction whatsoever.

This quick overview shows us that fines are the most common mechanism of punishing those who do not participate electorally. But one can also find a relatively high number of cases in which no sanction is imposed on them. For that reason, it is important to remember the distinction between formal and informal obligation to vote (Birch 2009). The former designates the actual enforcement of penalties on abstentionists, whereas the latter refers to social and/ or political pressure for an individual to cast a vote. Again, these are not mutually exclusive. But the distinction is important, as some authors who are in favor of compulsory voting argue that, even though formal penalties may be imposed, the determinant factor in driving individuals to polls will be informal sanctions (Hoffman & Graham 2006; Lijphart 1997). Their argument is that compulsory voting leads to the creation of an electoral participation culture, leading to the creation of a habit to vote and getting individuals to perceive abstention as morally wrong. However, empirical studies on the effects of compulsory voting suggest that its influence on turnout levels and, consequently, on turnout inequality, is only significant when there are formal sanctions (Blais, Massicotte & Dobrzynska 2003). On its own, moral obligation does not seem to be capable of raising electoral participation levels. But what is more, the probability of citizens to cast a vote seems to be higher when the penalties on abstentionists are heavier (Panagopoulos 2008; Singh 2011).

Nevertheless, some studies present results that do not seem to fit well into the ones I just referred. For example, in his study of the effect of compulsory voting on the level of political knowledge of citizens,

Sheppard (2015) concludes that it is on countries in which the fines on abstentionists are softer that individuals present higher levels of such knowledge. One can thus conclude that we still have a poor understanding of the effects of the harshness of penalties imposed upon abstentionists, given the small number of studies on that matter that have been published. It may be that such effects depend on the social, institutional and political characteristics of each country. Either way, it is clear that we need further studies on this matter, and that those studies need to take into account the specific characteristics of different countries. So far, all one seems to be able to state is that compulsory voting should, in fact, be accompanied by some kind of formal sanction on people who do not vote. But how harsh should those sanctions be, we cannot know for sure.

6. CONCLUSION

In this paper, I aimed at showing how compulsory voting can be a path towards reducing unequal political participation. I argued that, because it raises the levels of participation close to its maximum, it dilutes the differences between turnout levels of different social groups, in a way that no other measure can. However, I contended that the application of compulsory voting does not need to go against the application of other measures that aim at a greater involvement of citizens in the public matters of their country. In fact, I have argued that the chances of compulsory voting being able to achieve its beneficial potential are greater when its implementation is *accompanied* by such measures. Regarding the sanctions that should be imposed on abstentionists, I have shown that electoral behavior studies suggest that only formal sanctions seem to be effective in bringing about considerable increases in turnout levels and, thus, in fighting unequal political participation. However, it is hard to know just how harsh those sanctions should be, as the scarce empirical work on that matter presents some contradictory findings.

REFERENCES

- Birch, S. (2009). Full participation: A comparative study of compulsory voting. Tokyo: United Nations University Press.
- Blais, A. (2006). What affects voter turnout? *Annual Review of Political Science, 9,* 111-125.
- Blais, A. (2007). Turnout in elections. In R. J. Dalton, Russell J. and H. D. Klingemann (Eds.) Oxford handbook of political behavior. Oxford: Oxford University Press, 623-635.
- Blais, A., Massicotte, L., and Dobrzynska, A. (2003). Why is turnout higher in some countries than in others?. Retrieved November, 5th, 2015, from http://www.elections.ca/res/rec/part/tuh/TurnoutHigher.pdf.
- Blais, A. and Rubenson, D. (2013). The source of turnout decline. *Comparative Political Studies, 46(1),* 95-117.
- Brehm, J. W. (1999). "Reactance". In A. S. R. Manstead and M. Hewstone (Eds.) The Blackwell enciclopedia of social psychology. Oxford: Blackwell Publishers, 466-468.
- Burnham, W. D. (1987). The turnout problem. In A. J. Reichley (Ed.) Elections American Style. Washington, D. C.: Brookings Institution, 97-133.
- Fowler, A. (2011). Turnout matters: Evidence from compulsory voting in Australia. SRNN Working Paper.
- Franklin, M. (2004). Voter turnout and the dynamics of electoral competition in established democracies since 1945. Cambridge: Cambridge University Press.
- Gallego, A. (2014). Unequal political participation worldwide. New York: Cambridge University Press.
- Gratschew, M. (2004). Compulsory Voting In Western Europe. In R. L. Pintor and M. Gratschew (Eds.) Voter turnout in Western Europe since 1945: A global report. Stokholm: International Institute for Democracy and Electoral Assistance, 25-32.
- Hill, L. (2014). Compulsory voting defended. In J. Brennan and L. Hill (Eds.) Compulsory voting: For and against. New York: Cambridge University Press, 294-554.
- IDEA (2016). Voter turnout data for Greece. Retrieved February 25th, 2016, from http://www.idea.int/vt/countryview.cfm?id=89.
- Hoffman, J. and Graham, P. (2006). Introduction to political concepts. Essex: Pearson Education.
- Lever, A. (2009). Is compulsory voting justified? *Public Reason, 1(1),* 57-74.
- Lijphart, A. (1997). Unequal participation: democracy's unresolved dilemma. *The American Political Science Review, 91(1),* 1-14.
- Nie, N. H., Powell, G. B., and Prewitt, K. (1969). Social structure and political participation: Developmental relationships, part I. *The American Political Science Review, 63(2),* 361-378.
- Panagopoulos, C. (2008). The calculus of voting in compulsory voting systems. *Political Behavior, 30,* 455-467.
- Singh, S. (2011). How compelling is compulsory voting? A multilevel analysis of turnout. *Political Behavior, 33,* 95-111.

- Verba, S., Nie, N. H., and Kim, J. (1978). Participation and political equality: A seven-nation comparison. Chicago: University of Chicago Press.
- Verba, S., Schlozman, K. L., and Brady, H. E. (1995). Voice and equality: Civic voluntarism in American Politics. Cambridge: Cambridge University Press.

World Politics and International Relations

Sport for Development and Peace

Nfamara Sanneh
Diploma, Gambia Technical Training Institute, Gambia

1. INTRODUCTION

Sports universal popularity- As participants, spectators, or volunteers, people are attracted to sport arguably more than any other activity. This popularity transcends national, cultural, socio-economic and political boundaries and can be invoke with success in virtually any community in the world. Sports popularity derives in large part from the fact that, when done right, it is fun and enjoyable for everyone-participants and spectators alike. In context where people are faced with difficult and unrelenting challenges in their day to day lives, the value of this dimension of sport should not be underestimated.
Sports cross-cutting nature- Sport is one of the most cross cutting of all development and peace tools. It is increasingly being used to promote health and prevent disease, strengthen child and youth development and education, foster social inclusion, prevent conflict and build peace, foster gender equity, enhance inclusion of person with disabilities, and promote employment and economic development.

Sports potential to empower, motivate and inspire- Sport is inherently about drawing on, developing and showcasing people's strengths and capacities. By shining a light on what people can do, rather than what they cannot do, sport consistently empowers, motivates and inspires individuals and their communities in a way that promotes hope and a positive outlook for the future.

Sport as a communication platform. Over the past few decades, sport has emerge as global mass entertainment, and has become one of the most powerful and far reaching communications platforms in the world. Because global sport events offer the capacity to reach vast numbers of people worldwide, they are effective platforms for public education and social mobilization. By extension, high performance athletes have become global celebrities in their own right, enabling them to serve as powerful ambassadors, spokespeople and role models for development and peace initiatives.

'All great social change is impossible, until it happens, and then it was inevitable'. Albie Sachs, South African Supreme Court, Stuttgart.

2. HEALTH DEVELOPMENT AND SPORT

A healthy population is a necessary foundation for all development progress. Without healthy populations, the achievement of development objectives will be out of reach. Good health is fundamental to the ability of individuals to realize their full human potential. I t is also a crucially important economic asset. Low levels of health impede people's ability to work and earn a living for themselves and their families. When someone becomes ill, an entire family can become trapped in a downward spiral of lost income and high health-care cost. On a national scale, poor population health diminishes productivity and impedes economic growth, while investment in better health outcomes is generally seen as an investment in economic growth.
The close relationship between health and development is responsible for the importance given to health in MDGs: Health is the focus in three of the eight MDGs:

- MDG4: Reduce child mortality;
- MDG5: Improve maternal health; and
- MDG6: Combat HIV and AIDS, malaria and other diseases.

Health is a state of complete physical, mental and social well-being and not merely the absence of disease or infirmity. The enjoyment of the highest attainable standard of health is one of the fundamental rights of every human being without distinction of race, religion, political, belief, economic or social condition. Constitution of the World Health Organization, 1946.

Increasing physical activity levels- Because physical activity is a primary risk factor driving the global increase in chronic disease, sport can play a critical role in slowing the spread of chronic disease, reducing their social and economic burden, and saving lives. While physical activity includes a broader range of activities than sport alone, direct participation in sport is one of the most enjoyable, and therefore powerful, means of motivating and mobilizing people to become physical active. In addition to enhancing overall physical fitness, regular physical activity, active play and sports can have a positive impact on other major health risk factors, such as high blood pressure, high cholesterol, obesity, tobacco use and stress.

Fostering social connection- A powerful social connector, sport can bring people together, expand and strengthen social ties and networks, link people to resources and provide them with a sense of belonging. These social relationships are a fundamental determinant of health but are often lacking for people who are marginalized by poverty, disease, discrimination or conflict.

Sport can also be used to reduce the social stigma experienced by marginalized groups, such as persons with disabilities, people with HIV and AIDS, and former child combatants. By engaging these individuals in sport activities with other community members, sport creates a shared space and experience that helps break down negative perceptions and enables people to focus on what they have in common. This is an important step in enhancing these individuals 'self-concept and emotional health.

Promoting healthy attitudes and behaviors- In disadvantage communities and populations, where people are often define in terms of their needs and deficits, sport provides a powerful counter – balance to these perceptions. Participating in sport draws on people strength's and assets-energy, enthusiasm, natural and acquired skills, the desired to excel- and the universal capacity for fun and enjoyment. In this way, well designed sport programs that are inclusive, fair, and fun and promote excellence at all skill levels- help to empower participants and build self-esteem.

Self-esteem can be defined as a person's overall self-appraisal and feeling of self-worth. Self-esteem is critical to health because it motivates self-care and can contribute to health lifestyle behaviors. The acquisition of sport skills and life skills and friendship of others, the attention and guidance of coaches, and the examples set by them and other positive sport role models all encourage sport participants to believe in themselves, in others, and in their future. For individuals deeply affected by poverty, disease, disability or conflict, the development of self-esteem can be a profound psychological shift that enables and motivated them to adopt healthier lifestyle behaviors.

Sport as a platform for communication, education and social mobilization vehicle- Sport can play a valuable role as a communication, education and social mobilization vehicle. Sports entertainment appeal, amplified by global communications, has made it one of the most powerful communication platforms in the world. By engaging them and mobilizing high-profile elite athletes and professional sport clubs and federations, this communications power can be harnessed to deliver critical health information and messages, to model healthy lifestyle behaviors, and to marshal resources for health information and education, and can serve as a starting point for community mobilization to support health promotion, vaccination, and disease prevention and control effects.

3. SPORT, DEVELOPMENT, CHLDREN, AND YOUTH

Today's children and youth represent the single largest percentage of young people in history. There are 2.2 billion children and 1.5 billon youth in the world today. Of these, 1.9 billion and 1.3 billion respectively live in developing countries. Together, these young people represent an unprecedented opportunity to reduce global poverty and advance development.
In most developing countries today, young people make up the largest percentage of the population-in some cases more than 50%. In most cases, the number of young people will grow in the next 10-20 years. This means that countries will be facing significant economic pressures to fund secondary education and the prevention of non-communicable and infectious disease such as HIV and AIDS. The large number of young people also offers and historic opportunity. The emerging workforce is young, and the overall population contains relatively few elderly individuals and children to support. To maximize the opportunity this young cohort presents, it is important to invest in and support today's children and youth. Investing in children and youth today will ensure that they are healthy and well equipped to manage the critical life transitions ahead of them.

Sport as a tool to promote child and youth development- Sport can contribute significantly to international, national and local efforts to give children a healthy start. Sport can help those who haven't received a good start, and equip youth with the information, skills, personal and social resources, and support needed to make key life transitions successfully.

It is important to note, however, that much of the evidence supporting sports potential comes from developed countries. More focused research and evaluation of policies and programs is needed in developing nations, to generate knowledge and inform initiatives undertaken in these contexts. However, sport is already being use worldwide to advance child and youth development and education, suggesting that its benefits are already appreciated, if not yet fully understood or explained.

Existing evidence shows that developmentally appropriate programs for every young children can help ensure they receive the positive experiences and stimulation through play and establish a foundation for success,. Involving parents and caregivers allows programs to pass on simple and enjoyable ways for them to help children develop and grow through play.

'Children, after all, are not just adults in the making. They are people whose current needs and rights and experiences must be taken seriously'. Alfie Kohn.

4. GENDER DEVELOPMENT AND SPORT

Gender equity is a term use to describe both the principle and practice of fair and equitable allocation of resources to, and opportunities for, men and women. Gender equity eliminates discriminatory practices that are barriers to full participation for either gender. The practice of gender equity means changing responses to, and treatment of, men and women to ensure that gender is a neutral factor when accessing resources, rights, and opportunities.

Gender differences between men and women do not necessarily imply inequity. However, globally, women are particularly disadvantage by gender constructs which prevent them from fully realizing their rights, accessing resources, and harnessing opportunities.

Globally evidence of gender in equity includes the widespread preferences of sons over daughters, limited education and work opportunities for girls and women, and high levels of physical and sexual violence against girls and women. More subtle forms of discrimination are arguably equally problematic and include gender stereotyping and institutional discrimination. Some cultural traditions that favor men over women, and he subsequent cultural practices that perpetuate that discrimination, lead to exclusion. Gender inequity varies significantly, in both degree and in the forms it takes, from country to country. Some countries have made significant progress in reducing inequities through targeted actions to change gender norms, legislate and protect human rights, and ensure access to education health and other services.

4.1 Sport as a Tool to Promote Gender Equity and Empower Girls and Women

Sport is an integral part of the culture of almost every nation. However, it's used to promote gender equity and empower girls and a woman is often overlooked because sport is not universally perceived as a suitable or desirable pursuit for girls and women. Existing social constructs of masculinity and femininity- or socially accepted ways of expressing what it means to be a man or woman in a particular socio-cultural context-play a key role in determining access, levels of participation, and benefits from sport.

It is true in all countries that girls and women are less likely than boys and men to participate in sport, and sport continues to be dominated by males. It is a mistake, however, to assume that this is because girls and women do not wish to participate. Poverty, heavy domestic demands, safety concerns, lack of accessible transportation, inadequate sport and recreation facilities, and few opportunities for physical education and skill development frequently prevent women's participation in physical activity and sport.

'A transformed partnership based on equality between women and men is a condition for people-centered sustainable development.' Mission statement, Beijing platform for Action, Fourth United Nations World Conference on women, Beijing, 1995.

5. DISABILTY, DEVELOPMENT AND SPORT

Disability is any physical or mental condition that limits a person's movements, sense or activities. The term disability is conventionally used to refer to attributes that are severe enough to interfere with, or prevent, normal day to day activities. According to UN Convention on the Rights of Persons with Disabilities, 'persons with disabilities include those who have long term physical, mental, intellectual or sensory impairments which, in interaction with various barriers, may hinder their full and effective participation in society or an equal basis with others. Disabilities can be permanent, temporary, or episodic. They can affect people from birth, or be acquired later in life through injury or illness.

Sport as a tool for the inclusion of person with disabilities - Sport for persons with disabilities is not a new concept, but its full potential as a powerful, low cost means to foster greater inclusion and well-being for persons with disabilities is only beginning to be realized.

Sport works to improve the inclusion and well-being of persons with disabilities in two ways, by changing what countries think and feel about persons with and by changing what persons with disabilities think and feel about themselves. The first is necessary to reduce the stigma and discrimination associated with disability. The second empowers persons with disabilities so that they may recognize their own potential and advocate for changes in society to enable them to fully realize it. The community impact and individual impact of sport help reduce the isolation of persons with disabilities and integrate them more fully into community life.

Sport changes community perceptions of persons with disabilities by focusing attention on their abilities and moving their disability into the background. Through sport, persons without disabilities encounter persons with disabilities in a positive context (sometimes for the first time) and see them accomplish things they had previously thought impossible. Their assumptions about what happens with disabilities can and cannot do are profoundly challenged and reshaped by experience. As well, the tendency to see the disability instead of the person is greatly reduced, in part because of the common experienced of sport that they now share.

'No country can afford to turn its back on 10% of its population.' ILO/UNESCO/UNICEF/WHO proclamation, December 3, 1997.

Is China the Main Beneficiary in the Ukraine Crisis? The 'Eager Anticipation' and 'Rational Thinking'of the Sino-Russia Cooperation after the Ukraine Crisis

Xinli Guo
Student, Peking University, China

1. INTRODUCTION

Ever since the outbreak of the Ukraine Crisis, United States and the European countries have politically and economically isolated Russia and have established a united front against Russia. Due to the deterioration of the relationship between Russia and the western countries, Russia has turned its attention to its east, which consequently led to developing a closer bilateral relationship with China to curb its external and internal pressure. On May 2014, Putin declared Russia's support for China's 'One Belt One Road' initiative, and on April 14, 2015, the Russian government announced that it would join the Asian Infrastructure Investment Bank as a founding member state. During May 2015 when Chairman Xi Jinping made his visit to Russia, the two heads of state proposed for the first time to connect China's 'Silk Road Economic Belt' and Russia's 'Eurasian Economic Union'. From the economic aspect, Putin's visit to China and the signing of the 400 billion dollars natural gas supply agreement marked a monumental conclusion to the negotiation that lasted for almost ten years. On the aspect of security cooperation, Russia and China jointly held a military exercise to celebrate the victory of the 70th anniversary of the anti-fascism war. At the same time, Russia broke its habit and decided to provide the latest S-400 air defense missile system to China, which makes China become the first buyer of the equipment of such kind.

Such series of close cooperation between China and Russia have alarmed the western countries and much of the international society, leading them to be concerned about the newly developing rivalry. Following the enforcement of the ties between China and Russia after the Ukraine Crisis, the issue has gained considerable amount of coverage on media and the academia of international politics, some deliver a viewpoint that 'China is the only beneficiary from the Ukraine Crisis'[1]. Such statement clearly depicts the concern of the western society about China 'strategically benefitting' from the standoff between United States and Russia after the outbreak of the Ukraine Crisis.

This paper believes that the Ukraine Crisis has led to the deterioration of the relationship between the United States and Russia and has promoted closer ties between Russia and China. Also, there are reasons to believe that Ukrainian Crisis is the turning point of the development of the Sino-Russian strategic partnership. In the past relationship between China and Russia, China has always been in a passive position but China has taken a more active role in the relationship after the Ukraine Crisis.

While these developments are extremely positive, the fundamental problem of the contention between China and Russia still exists following the Ukraine Crisis. Under such conditions, the answers to 'how much impact Ukraine Crisis has on the relationship between China and Russia' and 'whether the crisis can fundamentally change the nature of the relationship between China and Russia' are actually quite clear Despite the nearing of the Russia and China may seem like a logical measure for the two countries, there are still a lot of uncertainties and vulnerabilities in the relationship between the two countries due to numerous historical and political variables. Ukraine cannot change the nature of the relationship between the two nations. The room for growth of Sino-Russian relationship is not as large as it is widely perceived to be.

2. UKRAINE CRISIS AS A CATALYZER FOR SINO-RUSSIAN COOPERATION

Historically speaking, there have been three periods of alliance and confrontation between Russia and China, and the extreme cases of alliance and confrontation have left scars for both of the countries that remain to this day. After the collapse of the Soviet Union, however, the relationship between China and Russia gradually further gained maturity and reason, and developed into the current bilateral relationship. It took a turn from the 'good neighbor-good partner' relationship of 1993 towards the 'the strategic partnership of co-operation oriented towards the 21st century' of 1996. The Sino-Russian Treaty of Friendship singed in 2001 replaced the Sino-Soviet Treaty of Friendship signed in 1950. 'To be friends forever and never be enemies' has become the highest standard for the relationship between Russia and China.

[1] *D Trenin, The Only Winner in Ukraine Is China, Moscow Times, August 20, 2014.*

According to the official statement, China and Russia are now the most important strategic partner to each other and the relationship prevail in their respective importance and particularity. The relation between the two countries is by no means a military or political alliance. It is a bilateral relationship that is accepted by the political leaders, the political elites, and the general public of both nations[2]. The maturation and the institutionalization of the political and economic relation between the two countries have provided social and legal grounds for the Sino-Russian partnership to face the instability of global politics after the Ukrainian Crisis.

2.1. The Ukrainian Crisis Strengthened the Mutual Trust between China and Russia

First of all, both China and Russia serve as the permanent member of the UN Security Council, sharing a common responsibility and value set on many international problems. On the international stage, both of them are eager to rewrite the rules deciding the global order, striving to shake the preexisting order under western dominance, which is evidenced by the consistent voting history of the two countries in the General Assembly. From 1974 to 2008, the Sino-Russian voting consistency was 82.53%, and it was 80.73% from 1992 to 2008. It is important to note that even in the 1970s and the early 1980s when China and the Soviet Union conflicted and kept distance with each other in diplomacy and military, the two countries still maintained their stance against the western countries. Thus, it can be seen that the two countries have always maintained a mutual understanding of each other's external policy regardless of the temporary rises and falls of the relationship.

Picture 1. The Voting of China, Russia (Soviet Union) and United States in General Assembly from 1974 to 2008[3]

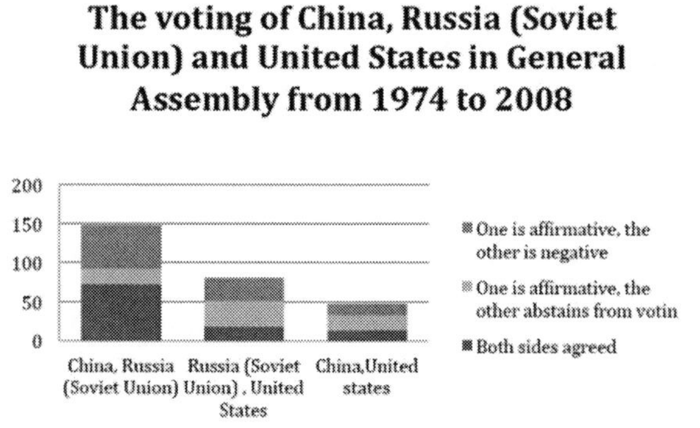

[2] *Guihai Guan, Cold Thinking on the Current Sino-Russian Relation[J].China Economic Report, 2013.*
[3] *Data source: United Nations General Assembly voting data.*

The Ukrainian Crisis further incentivized strategic mutual trust between China and Russia. Although both Russia and China are America's strategic rivals, Russia has long been yearning to restore its relationship with the west, thus regarded its relations with United Sates as a top priority. China has also put Sino-American relationship on top of its priority for the same reason. Thus, although China and Russia have kept in accord on many of the international issues, they still recognize that being in complete disagreement with United States and the western countries is not conducive for their national interest. The improvement of the Sino-Russian relation is in line with the diplomatic benefits of each other following the deterioration of the Russo-American relation, the launching of the economic sanctions of United States and Europe on Russia, the promotion of the Asian-Pacific rebalancing strategy by United States, and gradually intensifying conflicts between China and neighboring countries and United States on the maritime rights and interests surrounding East China Sea and South China Sea. During the special period when the Russian-American relation is sensitive and strained, China's politic stand on the Crimean issue or Russian President Putin's visit to China, the 'Conference on Interaction Confidence-Building Measures in Asia' held in Shanghai, the Sino-Russian '2014 Maritime Combined Military Exercises', and the conclusion of the agreement on gas supply between the two countries all represent that the Sino-Russian relation has entered into a new development stage.

Secondly, along with the growth of China's national power, China's sphere of influence in the international stage has expanded significantly. However, due to historical, cultural, and ideological differences, China has always been considered an outsider in the overall power structure dominated by the west. If you consider Russia, Russia has maintained its status as a powerful Christian nation and has wielded great amount of influence over the course of modern history. However, the political ideology of Russia has diverged with the rest of the western countries, reaching its peak during the Cold War. In addition, although over 20 years have passed since the collapse of Soviet Union, the cold war mentality is still restraining Russia from integrating into the western sets of values. This can be witnessed not only on a political level, but also from the cultural level, whereas Russia opposes Hollywood movies and western culture penetrating into its own[4]. Judging from the public media, the 'evil' image of Russia has further established itself

[4] *Steven Kurutz, "Russians: Still the Go To Bad Guys," New York Times, January 17, 2014.*

in the hearts of the public after the Ukraine Crisis. According to the opinion survey carried out by the Pew Research Center in 2014[5], the western region viewed Russia as unfavorable more than any other region, especially after Crimea was incorporated into Russia by force. As shown in figure 1, about 74% of the survey taken from Europe demonstrated that they have a negative impression on Russia. Figure 2 shows that the Europeans have a worse impression on Russia after Crimea was incorporated into Russia.

Figure 1. Median analysis of the Russian Unpopularity Degree in the World

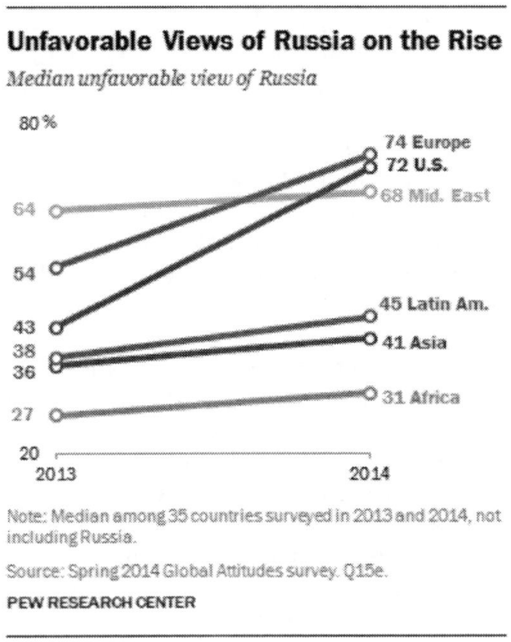

Figure 2. The Time Variation Analysis of the Russian Unpopularity Degree in the World

Europeans, Americans More Negative toward Russia

Do you have a favorable or unfavorable view of Russia?

	2013		2014		13-14 Change
	Fav	Unfav	Fav	Unfav	unfavorable
	%	%	%	%	
U.S.	37	43	19	72	+29
Poland	36	54	12	81	+27
UK	38	39	25	63	+24
Spain	38	51	18	74	+23
Germany	32	60	19	79	+19
Italy	31	56	20	74	+18
France	36	64	26	73	+9

[5] *Pew Research Center, "Russia's Global Image Negative Amid Crisis in Ukraine," July 2014, pp. 2-4.*

Then, in the current situation of global politics, it can be deduced that the cooperation of China and Russia is to prevent or reduce the unfair and unreasonable treatment of the western countries. Besides, considering the geopolitical factors and political ideologies, China and Russia have relatively similar sets of values, which encourage mutual trust and understanding. Ukrainian Crisis acted as a catalyzer to sever triangular ties between China, Russia and United States, providing China and Russia an opportunity to strategically cooperate and take a unified stance against United Sates.

2.2. The Improvement of the 'Warm Politics and Cold Economy' Status

The cooperation between China and Russia has always been promoted and implemented by the government with obvious policy characteristics. Thus, it is said that the Sino-Russian relation has always been in accord with 'warm politics and cold economy'. However, the Ukraine Crisis is a case where it promoted a large-scale trade cooperation between China and Russia, and to a certain extent made up for shortcoming of the 'warm politics and cold economy' between the two countries. Due to Ukraine Crisis, Russia was diplomatically and economically isolated by America and western countries and faced with their criticism, thus turning to China to break through the isolation. In the financial sector, important breakthrough has been made as trade settlements were signed with local currency and the payments were made through China's 'Union Pay' card system. The two sides have signed Agreement of the People's Bank of China and the Central Bank of Russia on the Bilateral Currency (RMB and Ruble) Swap Deal with a total amount of 150 billion RMB or 815 billion Rubles. In the energy sector, China and Russia jointly proposed the establishment of comprehensive energy partnership and the gas trade agreement after nearly ten years of negotiation: Mr. Putin made a visit to China and signed the gas supply agreement with a total amount of $400 billion, reaching consensus over the dispute focused on the gas price. The final price was set slightly lower than $380 per thousand cubic meters, the export price of Russia to European countries, and higher than $200 per thousand cubic meters, the import price of China importing from Central Asian countries such as Kazakhstan and Turkmenistan. The trade agreement, in the long run has political significance in addition to the economic reciprocity. The economic interdependence, particularly in the energy sector, has become one of the crucial interests to maintain the political relationships between the powerful nations with influence.

The sustainable growth of China further demands Russia's energy and raw materials, while the revival of Russia's status needs the funding and technology of China. China and Russia have complementary economic interests, which brings great deal of potential for cooperation. China is the country with the fastest economic growth not only in the Asia-Pacific region but also in the global level. It is logical for Russia to make use of the rapid development of Chinese economy to support its local economy in the far-east and coastal area. The bilateral cooperation through mutual interest has the potential to promote further cooperation of the two countries in political and military sector as well.

2.3. The Progress of the Cooperation in Local Economy

During the visit to Kazakhstan in September 2013, President Xi Jinping proposed the initiative to jointly establish the 'Silk Road Economic Belt'. Then, during his visit to Association of Southeast Asian Nations in October 2013, he put forward the idea to build the 'Marine Silk Road of the 21st Century'. The Silk Road Economic Belt and the Marine Silk Road of the 21st Century are referred to as the key components of the 'One Belt One Road' project.

For the One Belt One Road project passes through Russia's area of influence, thus geopolitically and economically, Russia is bound to play a pivotal role in the project and its cooperation is crucial for China. Originally, however, Russia stood against China's proposal for the One Belt One Road project. On May 29, 2014, under the lead of Russia, the leaders of the Republic of Belarus, Kazakhstan, and Russia signed the Eurasian Economic Union Treaty in Astana with the aim to establish a collective security organization and Eurasian customs and economic union to strengthen their military and economic cooperation. For Russia, this is to protect the region where its interest lay and to respond to NATO and EU's extrusion on Russia's national interest. The Silk Road Economic Belt crosses the region under Russia's influence, which makes Russia faced with numerous conundrums: first of all, China's economic cooperation with Russia's neighboring countries may bring damage to Russia's economic interests; then, China's strategic enterprising in the region may reduce the economic dependence of the countries in Central Asia on Russia, and give momentum to their diplomatic independence. Thus, the "One Belt One Road" project has failed to gain support from Russia when it was proposed.

However, the development of the Ukraine Crisis and the deterioration of the relationship between Russia and the west, the sharp fall of international oil price and the rapid depreciation of Ruble all have led to put serious internal and external pressure on Russia. Such pressure has led political elites and the critics in Russia to pivot their views to support China and its One Belt One Road project. In March 2015, the Russian government consequently issued a formal statement showing its support for the One Belt One Road project. The Russian first deputy Prime Minister Igor Shuvalov said in a speech that Russia believed the 'Silk Road Economic Belt' will bring opportunities for development to the Eurasian Economic Alliance rather than challenges. As a result, on May 8, 2015, President Xi Jinping and President Putin issued a joint statement on the cooperation between the Silk Road Economic Belt and the Eurasian Economic Alliance. The unique settings of the global politics that followed the Ukraine Crisis led the governments of the two countries to temporarily lay aside the sensitive issues between them such as free trade zones and turn towards cooperation to connection the two economic projects.

For China, the connection and cooperation between the Silk Road Economic Belt and Russia's Eurasian Economic Union has allowed One Belt One Road project to gain a major momentum. Russia is the key partner for the actual implementation of the 'one belt and one road' initiative: first of all, Russia itself is the unavoidable key node of the "one belt and one road" strategy. Secondly, Russia has irreplaceable right of speech in the central Asia, especially in he the Commonwealth of the Independent States.

It is worth noting that China's internal economic problems and its ambitions in international politics are the promotion of the Silk Road project. China's economic growth has slowed down significantly, and it is empirical for it to find a new source of impetus. The slowing of domestic consumption and excessive production has been considered as the two largest weaknesses of China's economy. After rounds of over-paced expansion of its steel, petrochemical, raw materials, and metals industry, they have reached or some have even exceeded the limits. The negative growth of PPI (Producer Price Index) over nearly 50 consecutive months and the negative corporate earnings clearly show that China's economic growth is slowing down. Faced with a new challenge, the rise of external demand followed by the One Belt One Road project has been proposed as a method to ameliorate the excess capacity. The economic growth of the Central Asian region has been limited due to the lack of infrastructure development in the region. In contrast, China has rich experience in the field of infrastructure construction, and a lot of infrastructure related domestic industries are in obvious overcapacity. The top five industries with excess capacity include steel, cement, electrolytic aluminum, glass, and shipbuilding industry, which are all crucial industries of infrastructure development. In addition, China over the course of its economic growth has accumulated a total of $4 trillion dollars of foreign exchange reserves with limited investment channels. The development of the infrastructure for the One Belt One Road project not only will drastically increase the expenditure of China's foreign exchange reserves, but also at the same time will improve China's overall economic circulation and the devaluation risk of the dollar reserves.

After the end of Cold War, although China has been trying to avoid untimely replacing Russia as the primary strategic rival of the United States, the United States still concentrates on its forces to firmly suppress the development of China. The intensifying of the situation in East China Sea and South China Sea together with the attitude of the United States towards the south China makes China extremely worry about the regional security situation in this region. The high-level Chinese government worries that the United States is likely to implement sea blockage on China and limit its maritime trade once serious crisis breaks out. Thus, China is in urgent need of a relatively safe strategic buffer and the westward movement which keeps away from the traditional influence sphere of the United States has become a wise choice. The Ukrainian crisis not only disturbs the plan of the United States to suppress China with the aid of Russia but also promote the cooperation between China and Russia. As a result, Russia cooperates with China to pin down and contend against the United States.

As for Russia, it is quite normal to have doubts on the 'one belt and one road', which is also in line with Chinese basic expectation on the great-power politics. Then, why did Russia change its attitude? This naturally is significantly promoted by the Ukrainian crisis. At the same time, Russia also has prejudged and reconsidered its own national interests. The first is out of the consideration for regional security. Russia has suffered great pressure from the west, whether from the eastern enlargement of NATO and EU or from the Crimean issue. To avoid being attacked front and rear, Russia needs a stable rear, and China which puts forward the 'one belt and one road' undoubtedly becomes the country which can provide the isolated Russia the largest support at present. This not only protects the security interests of Russia, but also determines the development degree of the current strategic partnership between China and Russia.

In addition, Russia also has incentives of its own to support One Belt One Road initiative. One is the development of its infrastructure, especially the construction of the transportation system. The 'one belt and one road' is conductive to the implementation of the Development Strategy of the Transportation Infrastructure of Russian Federation. According to the strategy, it is a main objective of the Russian transportation policy to establish a safe and unobstructed channel through Russia and to the rest of the world. Second is the development of the far eastern and Siberian region of Russia and China's assistance. Russia has previously stated that one of the motives for joining the Asian Infrastructure Investment Bank is to develop its far eastern and Siberian region. Thus, although the One Belt One Road initiative is likely to damage interests and influence of Russia in Central Asia, but the security and the long-term development interests overweigh the regional conflicts, let alone the arduous competition for its regional prestige.

3. THE CONTENTION BETWEEN CHINA AND RUSSIA STILL EXIST FOLLOWING THE UKRAINE CRISIS

The Ukrainian Crisis is undoubtedly a turning point for the strategic partnership between China and Russia. However, the Academia has different views on whether the close strategic partnership between China and Russia is a passive choice, or a long-term strategic move. Although the Chinese government has shown great expectation on the cooperation between China and Russia, yet many western scholars and even some Russian scholars believe that Russia and China are bound to compete for the one's own national interest due to lack of mutual trust and conflict of interests. They perceive that it is impossible for the two nations to maintain a long-term strategic partnership.

The author believes that the current Sino-Russian relation is neither as complicated as the western scholars describe it to be nor as hopeful as it is described in the policy blueprint. The Ukraine Crisis has certainly led to the improvement of the relationship between China and Russia, but it cannot change the nature of the relationship between the two nations. Due to the comprehensive influence of various historical and realistic factors, there are still a lot of uncertainties and vulnerabilities in the relationship between China and Russia. Regardless of the frequent exchanges between the political elites, the actual room for improvement of the relationship between China and Russia is not as large as it is perceived to be. The development of the strategic cooperation between China and Russia will also be limited by external variables from the global society.

3.1. Russia Lacks Recognition on China

The 'double-headed eagle', a metaphor commonly used to portray Russia's foreign policy after the Ukraine Crisis to benefit itself between the east and the west, is actually not accurate because Russia has the inherent impulse to reconcile with the west. As is expounded in the Conceptions of the Foreign Policies of the Russian Federation, the order of priorities of Russia's foreign policy is: the Commonwealth of the Independent States (The Republic of Belarus is the first), the European countries (the order is Britain, Germany, Italy and France), the United States, Asian countries (China, India, Japan, Southeast Asia, Iran, Korean Peninsula, Afghanistan), the Middle East, Africa, Central America and South America.

Although 80% of Russia's territory is located in Asia, its politics, economy, culture, and population is focused on Europe. Thus, both the tsarist Russia of the past and the current Russia have always had a demand to maintain its relationship with Europe. To integrate into the western society is the coveted goal of Russia. However, the realistic condition of global politics makes it difficult for Russia to do so. The western powers have always regarded the 'polar bear' as an alien, and responded to Russia with strong alert and exclusion. Although Russia has always encountered great difficulty in the process of returning back to Europe, it has continued to strive to restore its relationship with the west. At the same time, the Russia still maintains a sense of superiority over the east. It can be said that the recognition of Russia on Asia, including China, has been limited and practical. And it will maintain deep internal relationship with the western world no matter how serious the contradiction with the western world is. Historically, when Russia failed to expand to Europe, it chose the Eurasian ideology which often infringed upon China. As a result, it can be surmised that China will be used as a pawn to balance Russia's relations with America and the western world.

The outbreak of the Ukraine Crisis and Crimea forcefully returning to Russia has led Russia's relations with the west to reach a historic low point since the end of the Cold War. Because of the surge of anti-western sentiment in Russia and resent against the isolation and suppression of the western countries, Russia has begun to put more importance on its relations with the Asia-Pacific. However, this choice is a temporary and a passive one, and China is also conscious of this. Once the relation between Russia and the west is restored, Russia's need for maintaining partnership with China decrease significantly. Russia will then rethink the bilateral agreement reached during the crisis. As a result, some agreements may rise to contention or simply won't be kept.

3.2. The Unsolved Conflict of Geopolitical Interests between China and Russia

In history, with the decline of China, Russia has been eager to expand to the east. In particular, after 1750, Russia has forced the Chinese government to sign a total of 104 unequal treaties. During the period between 1689 and 1883, Russia had successively forced the Qing government to sign numerous treaties including Treaty of Nerchinsk, Qiaketu Treaty, Beijing Treaty, Treaty of Saint Petersburg, Khovd and Tarbagatai Treaty, consequently occupying over 400 square kilometers of China's territories and China's vassal state. Due to the historical disputes and grievances, the mutual trust between China and Russia still has great amount of room to fill regardless of the constantly improving bilateral relationship. As a result, deep anxiety and wariness has always existed amongst political elites and ordinary people of the two nations.

In fact, the border conflicts between China and Russia has always been the focus of the international relations academia. The discussion has never stopped even after China and Russia officially signed The Supplementary Agreement on the Eastern Section of the Boundary between China and Russia. It is worth noting that China's National People's Congress Standing Committee kept the agreement under a low profile, considering it to be 'traitorous'. The specific details are unknown to the public, and only a local newspaper in Beijing has mentioned it lightly. Although the border issue between China and Russia is resolved legally through negotiations and redrafting of the treaties, but it is still in doubt whether the agreement is recognized by the public. Some Chinese scholars believe that the biggest obstacle of the Sino-Russian relationship is the territorial disputes that remain unsolved. Meanwhile, some Russian scholars have taken advantage of the historical and geopolitical conflict to put forward radical opinions and politicize the issue. For example, Professor Andre Ivanov who serves in the Russia Altai Agriculture University once criticized the fact that the Siberian region was depicted in the Chinese history textbooks as a Chinese territory stolen by Russia, heating the border conflicts.

Moreover, in terms of geography, the far eastern region of Russia is in a vacuum state compared to the three northeastern provinces of China. The three northeastern provinces amount to a total area of 145 square kilometers, and a population of 120 million. The far eastern region of Russia has an area of 670 square kilometers, and a population of 6.2 million with less than 1 person per square kilometer. Although it is crucial to gain China's support for Russia's development of the far eastern region, the cooperation as remained under dispute. Russia has implemented a series of restrictive policies on China's participation in the development of the far eastern region, which hinders the region's development of the comprehensive cooperation between Russia and China.

As for the contention over Central Asia, although China has admitted Russia's interests and influence over Central Asia, and Russia has also realized the need for China's support for stabilize and develop in the region, but both nations are skeptical and contentious on who is to gain further influence in the region. As it can be seen from the events that followed during or after the SCO (Shanghai Cooperation Organization) summit in Ufa, where several proposals raised by China on The SCO Development Bank was not approved, the proposal to build the framework of the international road transportation agreement also failed to gain consensus. As for China's proposal to set up the Asian Infrastructure Bank, Russia only submitted the application close to the deadline

3.3. The China Threat Theory is Deeply Rooted

The 'China Threat Theory' is not only popular in western academia, but also has significant influence in Russia. Compared with most Chinese people who have a positive attitude towards Russia many Russian elites and citizens refuse to accept the transcendence of China. The senior Russian officials are doubtful about China's intention on Russia's far eastern region and Central Asia. Some scholars and officials have proposed to prevent Russia from leaning to China and becoming China's 'economic client' and 'provider of raw materials'. They also criticize that China is actually the biggest beneficiary of the Ukraine Crisis after all. [6]In February 2015, the director of Russian Council of International Affairs Andrei Kortunov said in an interview with the Russian media that all neighbors of China paid close attention to China's rise with anxiety and was skeptical of China's political and economic expansion. In response to the reporters' question *'What role does Russia play for the current Sino-Russian relationship?'* he answered: *'Russia may not be an older brother anymore. Russia is now an elder sister of China. You only need to respect an elder sister but need not entirely comply with her.'* This new formulation accurately shows the mixed feelings of Russia to China[7].

In fact, both Russia and China have a strong superpower mentality and big prowess. Neither China nor Russia is willing to become dependent to a great power. As seen from Russia's strategy after the Ukraine Crisis, Putin began to lead Russia to "look east", which included the active pursuit of developing the far eastern region and curbing the economic sanctions from the west. Although it is true that Russia's national policy is to promote the development of the far eastern region, the strategy will not necessarily lead to resolving of Sino-Russian border conflicts, nor will it result in sudden recognition of China from Russia's elites and its public.

3.4. The Imbalance of the Sino-Russian Economic Cooperation

Three imbalances of the Sino-Russian economic cooperation exist: First is the imbalance among the cooperation initiatives. The second is the disparity between the will of the central government officials and local government officials. The third is the objective disparity between economic cooperation and political relation.

The first is the imbalance between the cooperation initiative of China and Russia. Over the years, China has been holding a positive attitude towards the cooperation with Russia, however Russia remains reserved about the cooperation. The Joint Declaration Between China and Russia signed on May 8, 2015 states that the two nations will 'accelerate the construction of the cross-border traffic infrastructure such as the railway bridge connecting Tong Jiang and Leninsikuoye port, and the road bridge connecting Heihe and Blagoveshchensk port. Up to August 2015, the Chinese side has completed 60% of the Tongjiang-Leninsikuoye cross-border railway bridge, while the Russian side has not yet started the construction. This on the one side reflects the inefficiency of the bureaucracy existing in Russia and on the other side reflects Russia's mistrust of China. And in May 2014,

[6] *D Trenin, The Only Winner in Ukraine is China, Moscow Times, August 20, 2014*

[7] *А Кобзев: Сейчас Мы для Китая – 'Старшая Сестра'. EB/OL (2015/04/11) http://russiancouncil.ru/inner/?id_4=5317#top*

China and Russia signed gas contract with a total amount of $400 billion. This action is the passive measure taken by Russia to shift its strategic focus to the Asia-Pacific region specifically under the pressure of the sanctions from the United States and Europe. The director of the Moscow Strategic Assessment and Forecasting Center Sergei Grinyayev stated, 'The history of reaching the agreement shows that the Chinese are tough negotiators. It has been negotiated many years and was unsolved until we are in this difficult political situation'. Obviously, it is suggested that China has taken advantage of the collective sanction of the United States and Europe on Russia to obtain cheap energy supply, but the disparity among the attitude towards the cooperation certainly exist.

The second is the imbalance between the will of the central government officials and local officials. The most prominent feature of the Sino-Russian cooperation is that it is led by the central government. However, the vision between the governments is not necessarily in line with the vision of the local governments. The attitude of the central government and local government towards the economic and trade cooperation between China and Russia is often uneven. With both China and Russia having profound tradition of bureaucracy, the government often has low work efficiency and transparency. The federal government, state departments, and local governments often collide while carrying out the policies. Although they may agree with each other while creating the general strategy, they may conflict with each other on the specific implementations. In addition, the corrupt legal system and the fickle policies also have caused great obstacles to economic cooperation between China and Russia.

The third is the objective disparity between economic cooperation and political relation, namely the 'warm politics and cold economy'. All of the cooperation between China and Russia are based on economic needs, yet the scale of the economic cooperation between China and Russia is far less than expected between the two power nations. In 2013, the volume of China's foreign trade with Russia was only $87 billion, and the volume of trade between China and the United States in the same year was $520 billion. The volume of trade with Japan and South Korea respectively were $312.55 billion and $250 billion, and the volume with Europe was $560 billion. During the visit to Europe in 2014, President Xi stated that China hopes to increase the volume of trade with Europe to over $1 trillion. In 2014, the volume of Russia's foreign trade with Brazil was $90.2 billion, with Australia it was $130 billion, and with South Korea it was $270 billion. It is evident that the economic cooperation between Russia and China is still at a low level. In 2015, the total volume of the bilateral trade between China and Russia is 422.73 billion yuan with a decrease of 27.8%. Among it, the volume of export to Russia is 216.24 billion yuan with a decrease of 34.4% and the volume of the import from Russia is 206.49 billion yuan with a decrease of 19.1%.The trade surplus of 9.75 billion yuan with a decrease of 86.9%. Although China has made promise to carry out large-scale investment in Russia such as cooperation in oil and gas industry and construction of Moscow-Kazan high-speed rail, but the funding is lagged. Despite China and Russia have signed more than forty contracts covering energy, electricity, aviation and communication industries under the eyes of President Xi and President Putin on May 20, 2015, the real application has been below expectancy.

3.5. The Sino-Russian Relationship Remain under the Influence of United States

It is an objective fact that the Sino-Russian relationship is largely influenced by the Sino-American relationship and the Russo-American relationship. The influence of United States has always restricted the development of Sino-Russian strategic cooperation. Russia stretches across the central Eurasian region, and is in control of nuclear arsenal more threatening to United States than any other countries. Russia's geographic advantages and military strength have always been considered as a greatest threat to United States, and China as a rising power is also regarded as the rival of United States. Adhering to the Cold War mentality, United States has shown anxiety and concern over the development of the Sino-Russian cooperation. Thus, most of the time, United States adopts the hedging strategy of getting close to Russia and suppressing China. A few years ago, the United States carried out the 'Asia-Pacific rebalancing strategy', and laid its emphasis on controlling China. After the outbreak of the Ukraine Crisis, however, it directed its spearhead to Russia. Currently, United States is using its "two fists" to control both China and Russia, which is slowly showing as a policy beyond its reach. It is worthy to pay attention to how United States will adjust its foreign strategy in the future. There is no doubt that when United States directs its spearhead against China, China will suffer an economic and political pressure, and Russia will have less pressure as a result, and the current incentives of China and Russia to cooperate won't exist.

After the outbreak of the Ukraine Crisis, Russia's foreign policy has always been utilitarian. Although the crisis has promoted the development of the Sino-Russian relationship, but China should be reminded that it is only a temporary one. At present, whilst Russia is faced with the economic sanctions from the west and the interruption of the military technological trade with Europe. Thus, Russia not only needs international support in politics, but also needs to find an exporting country to sell its energy and weapons for foreign exchange, China best fitting as a partner. However, it can be surmised that the sanction of United States and on Russia will not continue for a long time because the sanction will only push the strategic cooperation between Russia and China. In the future, Russia will continue to seek balance between China and United States along adjustments to the changing international environment to maximize its national interests. Thus, Sino-Russian cooperation cannot hold as long as it is perceived to do so.

4. CONCLUSION

After the Ukraine Crisis, the argument that 'China is the only beneficiary of the crisis' has continued to appear in academia and political arena. Such view is based on the following two considerations: Firstly, when the western world is in conflict with Russia, Russia's export of oil and gas, and imports of agricultural products will rely more on China, which increases China's bargaining power. The trade between China and Russia is more likely to be settled in local currency, thus more Russian capital will be invested in China's financial system, which is be beneficiary for the internationalization of RMB and the China-funded banks. Secondly, from geopolitical perspective, the conflicts between the western world and Russia has an effect of reducing pressure of the west on China, and is beneficial for China to exercise its outward strategy.

Although the arguments above may seem reasonable, they still lack deeper research. Those who hold the first type of view have overlooked some important economic details. First of all, the Russian economy scale is limited, and is less than a quarter of China's GDP with grim growth prospects. In addition to the western sanctions, Russia is also in the face of economic stalling, capital flight, high inflation rate and shrinking domestic market. Russia is China's 10th largest trading partner, and ranks below Malaysia, Brazil and other countries above in order. Under this situation, it is unpractical to expect Russia to continuously and only rely on imports from China. Secondly, the RMB internationalization is a long-term and a natural process, not one that the sole influence of Russia can have a decisive role. The local currency settlement of Sino-Russian trade can be beneficiary for the RMB internationalization, but not absolutely required.

Then, does China have the opportunity to gain more power in geopolitics and international relations from its partnership with Russia? It is necessary to have a clear understanding that Russia after all is a 'double-headed eagle' more focused on relationship with Europe. Such principle will not change easily and United States will remain at the core of its foreign policies. Although the contradictions and conflicts between Russia and United States have continued to fuel Sino-Russian relationship after the Ukraine Crisis, it is still difficult for it to develop into a long-term partnership. The current trilateral relationship amongst China, Russia, and United States is unstable and uncertain.

Yet, further deterioration of the relationship between the western world and Russia does not coincide with China's interests. In the current era characterized by the highly interdependent global economy, a long-term, stable, and peaceful external environment is more important to China than a temporary and imbalanced trilateral relations or regional conflicts. The strife between the western world and Russia may exacerbate the regional instability, and deteriorated external environment is not conducive China's core strategies such as the One Belt One Road project. Thus, China may not truly wish to profit at other's expense, whether from the perspective of economic development or national security.

It should be admitted that Ukraine Crisis has brought unprecedented development opportunity for the cooperation between China and Russia, but the actual potential for the relationship between China and Russia and their strategic cooperation is limited. Therefore, it is necessary to think rationally about the Sino-Russian relation while having eager expectations. It may be wise for China to strike the iron while it is hot before it is too late.

REFERENCES

- As Russia and U.S. Struggle, China Rejoices, Moscow Times, Aug 24, 2014. http://www.themoscowtimes.com/article/asrussia-and-us-struggle-china-rejoices/505745.html.
- Trenin, D. (2014). The Only Winner in Ukraine Is China, Moscow Times.
- Podberezkin, A. and Podberezkina, O. (2015). The Silk Road Renaissance and New Potential of the

Russian-Chinese Partnership [J]. *China Quarterly of International Strategic Studies.*

- Guihai Guan (2013). Cold Thinking on the Current Sino-Russian Relation [J].*China Economic Report, 5,* 108-111.
- Yongquan Li (2014). The Great Game Reflected in the Ukrainian crisis [J]. *Russian Journal, 3,* 5-10.
- Yan Lu and Jinzi Qiu (2014). The Impact of Ukrainian Crisis on the Gas Cooperation between China and Russia and Central Asia [J]. *International Petroleum Economics, 22(9),* 1-3.
- Zhaocheng Zhou (2015). The International Public Opinion on the 'One Belt and One Road' Strategy [J]. *Social Sciences in Nanjing, 7,* 1-5.
- Xiujiao Li (2015). Analysis of the Experts of the Russian Idea Bank on the 'One belt and One Road' [J]. *Siberian Studies, 3,* 19-24.
- Xiaohui Li and Minghong Jiang (2015). The Impact of the Ukrainian Crisis on the Sino-Russian Relation [J]. *The Fortune Times, 3.*
- Xianju Wang (2014). The Impact of the Ukrainian Crisis on the External and Internal Policies of Russia [J]. *Russian Journal, 05,* 45-48.

Review of the Conflict Theories
An Analysis of Causes, Effect and Way ahead

Dr. Jagmeet Bawa
Assistant Professor, Joint Director, IRC-UHVE, Punjab Technical University, India

Harpreet Singh
Research scholar, IRC-UHVE, Punjab Technical University, India

ABSTRACT

The long history of humanity has witnessed a lot of progress. But despite, so much of advancements, innovations and developments, we can see that there are conflicts at international, national and local levels, which opens-up new issues of concern among the nations, states and societies. At International level nation-states are in conflict for different reasons. We have seen a number of cases of inter and intra regional conflicts within nation-states. We have conflicts among different provinces and so on. This study paper reviews and summarizes the research on the subject of conflict - its main causes and effects. Majority of the studies suggested that conflict is inherent in social change, conflict permeates each and every strand of human existence, conflicts are part of our everyday life; it is there in our close relationships, at societal level and at international level and most of the time it leads to violence. Different research studies defined that causes of conflicts are different, depending upon the positions, interests, goals, needs, and threats between parties in conflict, recognition, greed, self interests, security and ethnicity dilemmas. Most of the studies till date focused on the causes of conflict like regional, religious, ethnic, ideological and cultural. We find that not many studies have concentrated on root causes of conflict. There is need of identifying the main cause of conflicts all around us from an individual's perspective. Because when an individual is in conflict due to his beliefs and assumptions, leads to conflict in individual's behavior and work outside. So human being has to understand itself to evaluate his beliefs and assumptions which becomes the cause of conflicts later on. In this study paper we have identified the four levels of human living, in which we can see conflicts around us like conflicts in Individual, family, society and nature/ existence. An individuals contradiction in the self leads to conflicts in family, conflicts in society and conflict in nature / existence.

Keywords: Conflict, Individual, Beliefs, Assumptions, ethnicity.

1. OBJECTIVES AND METHODOLOGY

Our aim in this study has been to identify the 'root cause' and effects of conflicts. This study aims at firstly to review the literature available on the definitions, causes of conflicts and secondly to find out the levels of conflict from an individual's perspective, effects of conflicts at different levels of human living and main cause of conflicts.

We have adopted a multidimensional comparative approach encompassing theoretical considerations backed up by empirical evidence from political scientists, anthropologists, historians, authors of conflict study.

2. INTRODUCTION

Irrespective of so much of progress and development, we are facing conflicts in our day to day life. Science has enabled us to understand the reality all around us. Technology offered us newer ways of convenience. We have made so many advancements and innovations, but the question still remains – why we are in conflict? We can see conflicts at international level, national level, between states, between ethnic groups, societies and even at smaller interest groups. Even we can see there are conflicts at organizational level, where employees and employers are fighting for their respective objectives.

2.1 History of Conflicts

Different authors and philosophers have given different definitions of conflict. Since humans have always waged conflicts, humans have also always engaged in different ways to end them. Often, one side coercively imposes its will upon the other side, sometimes violently, and thus terminates a conflict. But the path of progress was not smooth; irrespective of so much of research in this field; conflicts obviously not abolished. Vladimir Ilyich Lenin (1870-1924) elaborated Marxism with his still influential analysis of the relationship between capitalism and imperialism, which generated conflicts

for radical societal transformations. Also, during this time, religious thoughts and practices were also developed in ways that proved relevant to conflicts at that time. Pacifist sentiments and commitments had long been a part of Christianity and other religions, often expressed by quiet withdrawal from worldly conflicts but it was not the proven method to avoid conflicts. The First World War (1914-1918) destroyed many millions of lives due to conflicts and also shattered, what seemed to have been illusions of international proletarian solidarity of global harmony from growing economic interdependence, and of rational political leadership. The establishment of the Permanent Court of International Justice (PCIJ-1922), the predecessor of the International Court of Justice, was provided for in the Covenant of the League of Nations. Laying the groundwork, between 1946 and 1969, many developments provided the materials with which contemporary conflict area was built. The evolutions of conflict study prevent future conflicts by building new transnational institutions and fostering reconciliation between former enemies. Globally, this was evident in the establishment of the United Nations (UN). The United Nations came into being in 1945, following the devastation of the Second World War. The UN Security Council has the primary responsibility for international peace and security. It has one central mission: the maintenance of international peace and security. The UN performs this by working to prevent conflict; helping the parties in conflict to make peace; peacekeeping; and creating the conditions to allow peace to hold and flourish. The International Court of Justice (ICJ) is the principal judicial organ of the United Nations (UN). It was established in June 1945 by the Charter of the United Nations and began work in April 1946.The Court's role is to settle, in accordance with international law, legal disputes submitted to it by States and to give advisory opinions on legal questions referred to it by authorized United Nations organs and specialized agencies. Regional organizations are, in a sense, international organizations (IOs), as they incorporate international membership and encompass geopolitical entities that operationally transcend a single nation state. They have been established to foster cooperation and political and economic integration or dialogue among states or entities within a restrictive geographical or geopolitical boundary. Examples of ROs include the African Union (AU), European Union (EU), the Caribbean Community (CARICOM), the Arab League (AL), Association of Southeast Asian Nations (ASEAN), South Asian Association for Regional Cooperation (SAARC), and Union of South American Nations (USAN). But still all these developments are not able to contribute much in limiting destructive international and domestic conflicts. So it is necessary to focus on identifying the root causes of conflicts, because up to now our main focus is on symptoms of conflicts, primarily. In order to establish'conflict free' societies we need to focus on the root causes of conflict, primarily.

3. MEANING OF CONFLICT

3.1 Definitions of Conflict

Ramsbotham[1], in his book defines conflict is universal feature of human society. Conflicts are the result of heterogeneity of interests, values and beliefs.

Morton Deutsch[2] described conflict is inherent in social change. Conflicts are necessary for societal change and transformation.

Albert[3] 1963, described conflict refers to consensual conflicts over interests, where disputants want same thing and dissensual conflicts over values, where disputants do not want same thing.

Meha dixit[4] 2004, mentioned conflict permeates each and every strand of human existence and often take shape of diabolic cyclic violence unless dealt with.

Bjarne et all[5] 2011, in their book described conflicts are part of our everyday life – they form in our relations, at societal level and at international level. Conflicts lead to aggregation, hostility and war. In case of conflict, the relations often turn scur and dialogue ends.

J.L. Steele[6] 1976, de scribed conflict once assumed to be inherent ingredient in human society. This assumption sprang from observations of raw animal nature.

Morton Deutsch[7] 1983, described conflict as difference in goals of parties involved in a given situation.

[1] *Ramsbotham, Introduction to Conflict Resolution: Concepts and definitions, 2011.*
[2] *M Deutsch, Conflict Resolution: Theory and Practice. Political Psychology, vol.4, No. 3, 1983.*
[3] *Ramsbotham, Introduction to Conflict Resolution: Concepts and definitions, 2011.*
[4] *Meha Dixit, Theories of Conflict Resolution: An Analysis. Institute of Peace and Conflict Studies, 2004.*
[5] *Ramsbotham, Introduction to Conflict Resolution: Concepts and definitions, 2011*
[6] *J L Steele, Conflict Resolution. Operational Research Quarterly(1970-1977), vol. 27, No. 1, Part 2, 1976, pp 221-30*
[7] *M Deutsch, Conflict Resolution: Theory and Practice. Political Psychology, vol.4, No. 3, 1983.*

Table 1: Definitions of Conflict

Author's name	Definition of conflict
Ramsbotham	Conflict is universal feature of human society,
Morton Deutsch	Conflict is inherent in social change
Albert	Conflict refers to consensual conflicts over interests, where disputants want same thing and dissensual conflicts over values, where disputants do not want same thing
Meha dixit	Conflict permeates each and every strand of human existence
Bjarne et all	Conflicts are part of our everyday life – they form in our relations, at societal level and at international level
J.L. Steele	Conflict once assumed to be inherent ingredient in human society. This assumption sprang from observations of raw animal nature.

The literature we have studied in this study paper is incomplete to define the conflict from Individual's perspective, because conflict is defined as contradictions in an individual give birth to conflicts which get expressed in the from of individual's behavior and work. If we see this whole existence, there is no conflict at all. So conflicts all around us are the outcome of contradictions in an Individual.

3.2 Causes of Conflict

There is different range of reasons for conflict. Conflicts can take its origins in economic differentiation, social change, psychological development, political organization.

Floyer Acland[8] described conflicts are the result of difference in positions, interests or goals, needs, fears between parties in conflict.

Roger Fisher[9] described conflict may arise from basic needs of identity, security, recognition or equal participation.

Bjarne Vestergaard et al[10] described conflicts are result of human interactions or issues on which people are disagree.

Laitin et al[11] described sources of conflict are identity formation, ethnic, religious, racial, tribal.

Crocker et al[12] concluded sources of conflict can be greed, self interest, security dilemmas, bad neighbors or neighborhoods and there is no apparent end in the sight of violence.

Conflict may be defined as a struggle or contest between people with opposing needs, beliefs, values or goals.

Galtung defined the contradictions, difference in attitudes and behavior give rise to conflicts.

The literature we have studied in this study paper is incomplete to list out main cause of conflicts from an individual's perspective. The contradiction of beliefs and assumptions in an individual is the root cause of conflicts. The behavior and work of an individual is the direct outcome of beliefs and assumptions in an individual. The beliefs and assumptions without verification keep on changing w.r.t. time and place, which leads to contradictions in an individual. Contradictions in an individual expressed outside in an individual's behavior with other human beings and work with rest of nature.

4. LEVELS OF CONFLICT

In this study we are seeing conflicts from an individual's perspective. So whenever we see an individual lives in this existence at a time, lives at four levels of his living. If we look at our living, it becomes clear that an individual's living can be seen as encompassing at the four levels like - an Individual, Family, Society and Nature/Existence as figure 1 show. Whenever an individual lives, he lives at the level of individual, at the level of family, at the level of society and at the level of nature/existence.

Figure 1. Levels of an Individual's Living

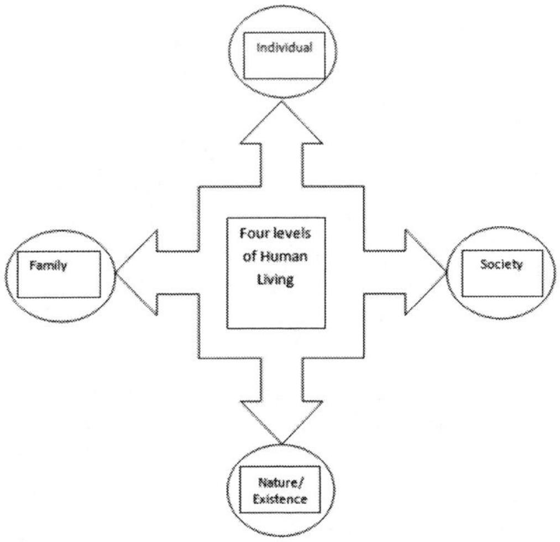

[8] *Ramsbotham, Introduction to Conflict Resolution: Concepts and definitions, 2011.*
[9] *R Fisher, W Ury, and B Patton, Getting to yes: Negotiating agreement without giving in. New York: Penguin Books, 1991.*
[10] *Website: http://lnu.se/polopoly_fs/1.105781!2011%20DCCR_BASIC%20MATERIAL.pdf*
[11] *E Babbitt and F O Hampson, Conflict Resolution as a Field of Inquiry: Practice Informing Theory, 2011.*

[12] *Website : http://fletcher.tufts.edu/~/media/Fletcher/News%20 Images/Babbitt.pdf*

Table 2: Causes of Conflict

Author's name	**Floyer Acland**	**Roger Fisher**	**Bjarne Vestergaard et al**	**Latin et al**	**Crocker et al**	**Galtung**
Causes of conflict	difference in positions	basic needs of identity	human interactions	identity formation	greed	contradictions
	interests	insecurity	issues	ethnic	self interest	attitudes
	goals	recognition		religious	security dilemmas	behavior
	Needs	equal participation		racial	neighborhoods	
	Fears			tribal		
	Threats					

Each one of us lives at all these levels, whether we are aware of it or not. An Individual is not able to understand itself, so he is in conflict with all the four levels. We can easily see the conflicts of an individual at different four levels all around us like conflicts with own self, conflicts in families, conflicts in society and conflicts in nature/existence.

5. EFFECTS OF CONFLICT

On looking into the different levels of an Individual, it becomes evident that irrespective of so much advancements and innovations, we still are in conflict at all the four levels of our living as described below:

i) At the level of individual - We can see that there continue to be issues in individuals with respect to lack of clarity on life goals, contradictions and stress. Purposelessness or dis-clarity about what to do?, disinterest in learning, lack of curiosity, boredom, escape to television and gaming, self-centeredness and disregard for others, inactive lifestyles, overweight and other health problems, substance abuse and risk taking, depression, suicide, stress, insecurity, loneliness and psychological disorders. An individual's behavior depends upon his beliefs and assumptions. Beliefs and assumptions without verification keep on changing. This leads to contradiction between belief and assumptions. It becomes source of conflict in an individual, due to which we can see above said effects of conflict at the level of an individual.

ii) At the level of family – Group of individual forms a family. So if individuals are in conflict with each other, families are also in conflict. We can see problems related to mutual understanding, increasing events of mistrust, insecurity and generation gap. The all consuming pursuit of many by any means, peer pressure, competition, shrinking circle of families; to be cared for-while exploiting every other, lack of communication with siblings, breaking of families, mistrust and insecurity in relationships, divorce, family feuds, incest, legal suits.

iii) At the level of society – Group of families takes the form of society. We can see easily increase of communal conflicts, exploitation of people and strife, terrorism effecting nations and violence in various forms. The increasing problem of corruption and other unethical practices in various fields of profession is also giving birth to conflict at various fields. Teasing, ragging, little engagement with social systems, feeding of mistrust in system, migration of people from villages to metro cities, men exploiting men, corruption in government, growing incidences of violence, terrorism, rising communalism, spreading casteism, racial and ethnic incidences of struggle, attempt of genocide and fear of nuclear, genetic and biological warfare.

iv) At the level of nature – Inevitable destruction of nature, non-evolvement with nature, cruelty to animals, climate change, water scarcity and pollution of air/soil, noise pollution, resource depletion of minerals and mineral oils, sizeable deforestation and loss of fertility of soil. We have increasing problems of environmental degradation in various forms as well as resource depletion threatening the very survival of human race. In fact all these problems are direct outcome of the well planned human activity carried out in the name of progress or development and in spite of all advancements of science and technology. This is the most important question intriguing human mind today.

We can see all these kinds of conflicts all around us at the four levels of our living at current state. But if we explore this nature comprising of materials, plants, animals and human beings, there is provision

Figure 2. Effects of Conflicts at Various Levels of an Individual's Living

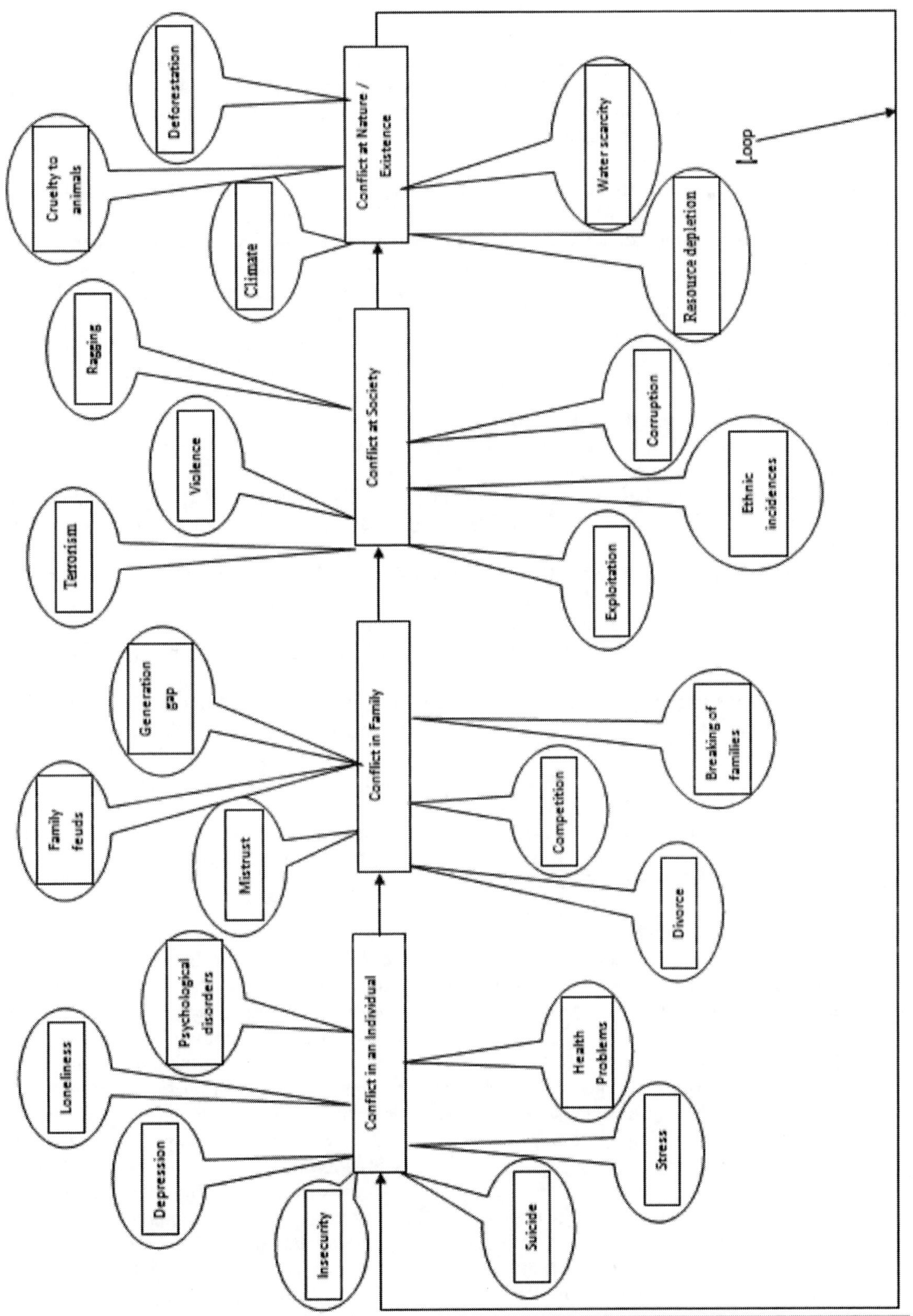

of harmony in nature, nothing is in conflict in the nature. Everything in this nature is in the form of rule, control and balance. We can see and verify in ourselves that basic cause of conflict is, an individual is having contradictions of beliefs and assumptions. If an individual is in contradictions in the own self, it gives raise to conflicts in family, which escalates conflicts in society, which is the reason of conflicts of an individual in nature/existence. Conflicts in nature again give birth to conflict in an individual as shown in the diagram; there is loop from conflicts in nature/existence to conflicts in an individual called vicious circle conflict loop.

6. ROOT CAUSE OF CONFLICTS

In this study paper we are focusing on an individual as a cause of conflict. If we don't deal with issues at hand, solutions will be short lived. So we have to identify the root cause of conflicts in an individual. The behavior of an individual with other human beings depends upon the beliefs and assumptions in an individual's self. Nowadays, an individual is not evaluating his beliefs or assumptions and he is leading his life on the basis of these unverified assumptions and beliefs. Generally is labeled as individual's personal life and freedom. We normally become very sticky about them without really evaluating them. When these come in conflict, we try hard to search out justifications and we generally put efforts to defend our beliefs or assumptions.

Beliefs and assumptions in an individual without verification lead to conflicts in the self of an individual. Beliefs and assumptions without verification may or may not be true. So we are not sure of what we believe because we have assumed these things true and have not verified them for ourselves. When an individual start living with other human beings, with contradictions, conflicts get reflected in his behavior. Because we have not verified these on our own behalf in our living with people and rest of nature, so we are not sure of their results. In fact it is quite obvious, since we find today that while sometimes we may have fulfilling relationships with people, we may be in conflict at particular point of time; sometimes we contribute to enriching the nature but mostly we are in conflict with nature.

In this study we have tried to discuss the main causes of conflict from an individual's perspective. Because till the time being we are not able to find out the root causes of conflict and all these efforts for the solution will not be long lasting. Because right now most of research is going on to resolve the conflicts, most of studies are work on the symptoms, but a little effort is for identifying the root cause of conflicts. So the need of hour is to seriously list out main causes of conflict to actually resolve the conflicts. If we don't deal with issues at hand the solutions will be very short lived and if we don't deal with all the four levels of an individual's living the solution will not be universal.

An individual is not verifying the beliefs and assumptions on the basis of what is naturally accepted to him. It leads to contradictions in an individual. When there are contradictions in individual, he is not able to understand the rest of the existence (i.e. everything that exists) and his relationship with all the levels of existence. So that's why an individual is in conflict with others. The beliefs and assumptions without verification is the reason of contradictions in an individual. The contradictions in an individual are expressed outside in the form of his behavior with other human beings and in the form of work with rest of nature. Such a conflict, always starts in an individual and it is the individual only which transforms this conflict to family, society and nature.

When beliefs and assumptions get evaluated by individuals, these get transformed to universal values. Thus evaluation of beliefs and assumptions in an individual on the basis of what is naturally accepted to an individual is the transformation in an individual. This process transforms these beliefs and assumptions into universal values. It will resolve all the contradictions in individual. And if an individual has resolution, he can live without conflict, which reflects in his behavior with other human beings of mutual happiness and through mutual fulfilling work with rest of nature.

7. CONCLUSION

We can see the conflicts at international levels, national levels, and local levels between states, districts and in organizations. In the present study we have discussed the main causes of conflict and their effects, we can see all around us. We have tried to see the cause of conflicts from an individual's perspective. Individual is an actor for all the activities at all the levels. An individual always lives at four levels like as an individual, family, society and nature/existence. An individual is in conflicts at all the levels of living. Now a days, we are studying all the external factors but not the individual who is creating all the conflicts and the one who is working for the resolution of all the conflicts. The main cause of conflict in an individual is the beliefs and assumptions without verification on the basis of what is naturally accepted to an individual leads to contradictions in an individual. Contradictions in an individual are expressed outside in the form of behavior with other human beings and in the form of work with rest of nature. Contradiction in an individual leads to conflicts in the family and which escalates the conflicts in society and finally becomes source of conflicts in nature/existence. The main cause of all kinds of conflicts in an individual is contradictions in an individual on the basis of wrong beliefs and assumptions. When these beliefs and assumptions get evaluated gets transformed into universal values. This transformation in an individual resolves the contradictions in an individual, which leads to harmony in an individual. Now when an individual behave with other human beings leads to mutual happiness and work of human being with rest of nature will lead to mutual fulfillment. Thus understanding in an individual can eliminate all the contradictions and can resolve the conflicts between human beings. The further scope of this present study is to develop the process of evaluating the beliefs and assumptions in an individual, because beliefs and assumptions should be evaluated from universal aspect, so that these get transformed to universal values and what should be the universal and holistic solution of conflicts at different levels of an individual's living from individual, family, society to nature, if an individual is the soul cause of conflicts.

REFERENCES

- Ramsbotham (2011). *Introduction to Conflict Resolution: Concepts and definitions.*
- Deutsch, M. (1983). Conflict Resolution: Theory and Practice. *Political Psychology, vol.4, No. 3.*
- Ramsbotham (2011). *Introduction to Conflict Resolution: Concepts and definitions.*
- Dixit, M. (2004). *Theories of Conflict Resolution: An Analysis.* Institute of Peace and Conflict Studies.
- Ramsbotham (2011). *Introduction to Conflict Resolution: Concepts and definitions.*
- Steele, J.L. (1976). Conflict Resolution. *Operational Research Quarterly* (1970-1977), *vol. 27, No. 1, Part 2,* 221-230.
- Deutsch, M. (1983). Conflict Resolution: Theory and Practice. *Political Psychology, vol.4, No. 3.*
- Ramsbotham (2011). *Introduction to Conflict Resolution: Concepts and definitions.*
- Fisher, R., Ury, W., and Patton, B. (1991). *Getting to yes: Negotiating agreement without giving in.* New York: Penguin Books.
- Website: http://lnu.se/polopoly_fs/1.105781!2011%20DCCR_BASIC%20MATERIAL.pdf
- Babbitt,E. and Hampson, F.O. (2011). *Conflict Resolution as a Field of Inquiry: Practice Informing Theory.*
- Website: http://fletcher.tufts.edu/~/media/Fletcher/News%20Images/Babbitt.pdf
- Marshall, Monty G., and Gurr, T.R. (2005). *Peace and Conflict.* College Park, MD.: Center for International Development and Conflict Management, University of Maryland.
- Rouhana, N.N. (2000). *Interactive Conflict Resolution: Issues in Theory, Methodology and Evaluation. In International Conflict Resolution after the Cold War,* edited by Paul C. Stern, and Daniel Druckman. Washington, D.C.: National Academy of Sciences.
- Volkan, V. (1998). *Bloodlines: From Ethnic Pride To Ethnic Terrorism.* New York: Basic Books.

Social Affairs and Medicine

Gender Equality in Saudi Arabia: New Steps, New Hopes

Nouf Ahmed Al Yousef
Student, Al Yamamah University, Saudi Arabia

ABSTRACT

The purpose of writing this research is to highlight the initiatives of Saudi Arabia in the past decade in regards to gender equality. We can notice as Saudi women the progress that the country has made in empowering women. Starting with the development of the education programs, building the largest women's university in the world, representing Saudi Arabia in international events and having the right to vote in country's municipal elections like men do. We cannot dismiss the recognitions and awards that have been given to Saudi female researchers by our country as an appreciation of their achievements in raising the quality of medical fields. However, we cannot ignore the fact that Saudi women are still facing several difficulties and challenges in the workplace, such as having unequal payments as men do.

On the other hand, when King Abdullah bin Abdulaziz endorsed women to enter Al Shoura Council for the first time by an increase by 20 percent in seats, which is a total of 30 women out of 150 members, this decree opened infinite opportunities to Saudi women locally and internationally. Indeed, if we compare the country's status 20 years ago to today, we would never think that we would have that amount of improvements regarding women's matters.

ITRODUCTION

Saudi women are limited in terms of rights when compared to its neighbors. In 2013, the Global Gender Report released by the World Economic Forum placed Saudi Arabia at position 127 among 136 nations around the world for steps made to realize gender parity (Al Arabiya 2013). To date, women irrespective of age must have male guardians in terms of government transactions and the country is the only one in the globe that still forbids women to drive. With regard to the workforce, it is noteworthy that they constitute only 13%.

Notwithstanding, this apparent gender discrimination, the status of women in Saudi Arabia in the recent times has outstanding improvements. Most important was King Abdullah's decree in 2011 allowing women to vote in the 2015 local elections in addition to appointment to the nation's Al Shoura Council. It is, however, important to note that despite the limited social, economic as well as political opportunities, women have achieved much in the education realm. In this respect, Saudi women account for the largest component of graduates and their literacy level is 91% (Al Arabiya 2013). These figures are remarkable considering that it has taken only four decades for the conservative nation to realize parity in this regard. Higher education and research, leadership and decision making, the business sector and the workplace and iconic Saudi women are topics that will be discussed in this paper in details.

1. HIGHER EDUCATION & RESEARCH

1.1. Higher Education

Despite the protracted as well as remarkable array of denied civil liberties, women in Saudi Arabia have made significant contributions and strides in the important field in higher education. Women ability to access universities is laying the ground for equal rights in the future as they seek to acquire more opportunities in employment. Women have in this regard made systematic efforts in the educational sector, especially in higher education, to improve their conditions within the society by focusing on the constructive role women can play in political, social and economic facets in the country (Oxford Business Group 2014).

Moreover, the famous quote by King Abdullah that motivated Saudi female and has given them a light of hope says *'I believe strongly in the rights of women. My mother is a woman. My sister is a woman. My daughter is a woman. And my wife is a woman'* (Arab News 2012). Correspondingly, amongst the programs that the government had established to support the education level in Saudi Arabia, is the King Abdullah Foreign Scholarship Program that serves male and female Saudi students equally to study abroad. Keep in mind that the expenses are fully covered by the country; hence, offering scholarships will hopefully lead the next generation to strive FOR nothing but the best. Moreover, this decree allows Saudi females to have an equal world-class education by getting their bachelor, master and PhD degree without any barriers stopping them from proceeding their learning abroad. Consequently, the positive outcome that resulted from that decree is having 150,109 males and 49,176 females' graduates– roughly a 3:1 ratio (University World News 2015).

However, women's role remains limited as they are not as yet allowed to make significant decisions in the education sector, meaning in that they have more opportunities in education than ever before, but they have not been accorded the chance to realign the sector to meet their specific needs as well as desires (Cordesman 2013).

Through King Abdullah's initiatives, the Saudi government has continuously implemented policy modifications in the education sector especially higher education. The country has constructed the largest all women's university in the world to improve the outcomes of women in the education sector. Currently, Princess Noura bint Abdul Rahman University has 60,000 students attaining bachelor degrees, and undergraduate as well as post-graduate courses. The government has also appointed a woman as the assistant minister for education to ensure the realization of women empowerment through education (Al Arabiya 2013).

1.2. Research

Research wise, Saudi females have done several important achievements that reflect their limitless capabilities toward learning. For example, Prof. Ghada Al Mutairi specialized in Nanomedicine, a professor at University of California; she came up with a light therapy treatment using Nano to target disease without affecting the whole body. When we mentioned Saudi women researchers, we cannot forget Dr.Khawla Al Kuraya'a a Saudi physician and cancer specialist. She is a professor of pathology and directs the King Fahad National Center for Children's Cancer and Research. Al Kuraya'a is the first Saudi woman who received Order King Abdulaziz Al Saud award and appointed by King Abdullah bin Abdelaziz to be one of the Al Shoura members too. Moving from cancer to the blood bank is where Dr.Hind Al Humaidan is the first Saudi who is specialized in blood diseases. She was chosen to be the first female director of Blood Bank and Transfusion Services and is the Consultant Hematopathologist at the King Faisal Specialist Hospital and Research Center in Riyadh. Her achievements are remarkable because she introduced a new field of Medicine to Saudi Arabia, which is Stem Cells. She have been working for 27 years at King Faisal Hospital.

Figure 1. Dr. Hind Al Humidan

Figure 2. Prof. Ghada Al Mutairi

2. LEADERSHIP & DECISION MAKING

2.1 Al Shoura Council

King Abdullah bin Abdelaziz had a vision toward a bright future for Saudi women. To begin with, King Abdullah bin Abdullaziz in 2011 made a momentous decree permitting women to participate in the traditionally all-male Consultative Assembly. The decree made amendments to the articles rules by introducing a quota of (20%) of Al-Shoura members of women which means 30 out of 150 members of Al-Shura Council should be represented by women. (Al Arabiya 2013). Secondly, in 2014 the Shoura Council approved a draft statute on violence meted out against women for forward submission to the King. This law defines different protection aspects against abuse and legal measures against accused persons. It also seeks to increase public awareness about gender-based violence and its implications for women and society.
Al Shoura has, together with the Ministry of Justice, spearheaded a campaign since last year to allow women practice law in courts. This is a great milestone considering that women were not allowed to practice as lawyers even though the law allowed them to

study law from 2006 thus affording them the chance to tackle social challenges by leveraging judicial reforms. Furthermore, Al Shoura has empowered Saudi women by ensuring that their participation in the assembly is not meant to gain political expediency (Al Arabiya 2013).

Finally, the changes introduced by Al Shoura to empower women are not exclusive to the political front but also include the social one.

2.2 Key Government Positions

Saudi government believed in women's managerial skills. In fact in 2011 King Abudallah bin Abdulaziz declared hiring Dr.Huda Al Ameel President of Princess Noura bint Abdulrahamn University as the largest only Women University in the world. Al Ameel was qualified enough to prove herself by being the president of PNU for 5 years, leading more than 60,000 students. (SPA 2011).

2.3 Municipal Elections

For the first time in history, Saudi women are allowed to vote in Municipal Elections. Moreover, based on King Salman bin Abdulaziz decree Saudi female were honored to join the process of registration for the Municipal Elections In Aug 2015.That decree was a continues of what King Abduallah bin Abdulaziz has started regarding women's equality. The voting periods has ended with 900 women who voted for their seats and 18 women were chosen (Yosri 2015).

3. THE BUSINESS SECTOR & THE WORKPLACE

3.1 Women Employment

Nowadays, the challenges facing Saudi women in the workplace are diminishing by the day. With the support of the king and the government, the percentage of females occupying managerial positions especially in private sector ,leading big teams, and even managing their own companies has increased significantly the past two decades. Moreover, the government has supported and implemented several laws to support the role of women in the workplace. For example, in private companies, a Saudi woman equals two foreign employees, meaning that in order for a private company to get two foreign visas they must employ a Saudi female. Additionally, several male managers confirm that women are more dedicated and more passionate in the workplace than men, which means that females are capable and able as males to reach higher stages in their careers.

On the other hand, Saudi women are still struggling at the workplace in having equal rights. For Example, in some cases women are receiving few salaries than men occupying the same exact position and job title. In the public sector, women are receiving different and inflexible treatment where they do not have the opportunity to go for training programs abroad or occupying managerial positions. For instance, it is very common to see men employees in public institutions attending training programs abroad.

3.2 Women Entrepreneurs

The number of entrepreneurs in Saudi Arabia has increased significantly the past decade, which is in fact a benefit for the development of the country's economy. Opportunities are more available for women compared to 20 years ago, where being a woman entrepreneur was an unusual topic. On the other hand, private and governmental funding programs are more accessible for entrepreneurs to start their businesses. For example, Princess Reema bint Bandr is the owner of the luxury retail corporation Alfa International Riyadh and Alf Khair Academy for supporting women's activity is one of the top female entrepreneurs in Saudi Arabia. Moreover, her latest achievement was introducing the breast cancer campaign '10K' by having the world biggest breast cancer ribbon with 10,000 women who participated. It was approved by Guinness world records (Diamond 2015).

4. ICONIC SAUDI WOMEN

Saudi women are recognized not just nationally but also internationally. Thoraya Ahmed Obaid is the first Saudi woman who worked in the UN in 1975 and had the chance to be the first Saudi woman to receive the government scholarship to study abroad. Obaid, served 25 years at United Nations to empower women and support gender equality. 'UN Secretary-General Kofi Annan has described her as the ideal candidate to head the UNFPA, an agency dedicated to promote equality between the sexes as well as universal education and health care, especially better reproductive health care for women' (Tao, N.d). Moreover, In 2013 Thorya Obaid was chosen by King Abduallah bin Abdulaziz to be a member and the Head of Human Rights Committee at Al Shoura Council (Crossette 2013).

Furthermore, in 2012, for the first time, the Kingdom of Saudi Arabia allowed women to participate in a major athletics completion-the Olympics, when Sarah Attar and Wojdan Shahrkhani participated in the 800m race and judo respectively. This marks a major achievement for Shoura Council in improving the status of women, and is a major leap forward for Saudi women in achieving gender parity (Oxford Business Group 2014).

Figure 3. Thoraya Ahmed Obaid

Figure 4. Sarah Attar

CONCLUSION

In conclusion, despite the improvements that Saudi Arabia have made in the direction of empowering women, there is still much more that needs to be accomplished to increase women's employment rate and answer their needs in regards to work and education. Saudi women have proven themselves in unique felids nationally and internationally. Moreover, building Princess Nourah University was a turnover for the education sector for females in Saudi. In addition, the royal decree that King Abduallah had made allowing females to be members of Al Shoura council was a big step towards empowering women in Saudi Arabia. However, obstacles are still facing Saudi females in workplace in many different areas such as training programs and promotions, which needs to be taken care of.

REFERENCES

- Al Arabiya (2013, January 11th). Breakthrough in Saudi Arabia: Women allowed in Parliament. Retrieved December 12th, 2015, from Al Arabiya News: www.alarabiya.net/artices/2013/01/11
- Cordesman, A.H. (2013). Saudi Arabia Enters the 21st Century. Greenwood Publishing Group.
- Oxford Business Group (2014). Saudi Women's Role in Improving Education. Oxford Business Group.
- Adah Almutairi. The Art of Falling Apart. benefunder: http://benefunder.org/sites/default/files/content/cause/picture/Headshot_20.jpg.
- KIM KYUNG. Wojdan Shaherkani (يلع نادجو اينأخرهش نيدلا جارس) ,Judo , FIRST SAUDI WOMAN TO COMPETE IN OLYMPICS , London 2012 Olympics . 柔道, 首次沙烏地阿拉伯女性在奧運競爭 2012, 倫敦奧運. نم دادرتسالا مت est100: • https://www.washingtonpost.com/blogs/london-2012-olympics/files/2012/08/saudi2012-08-03T094256Z_01_OLYKR682_RTRIDSP_3_OLY-JUDO-JUW79K-ROUNDS-DAY7.jpg.
- Manail Anis Ahmed (2015,December 11th). Saudi students abroad. University World News: http://www.universityworldnews.com/article.php?story=20151208190817896.
- RIMA AL-MUKHTAR (2012, May 6th). Empowering women: Kingdom leads the way. Arab News: http://www.arabnews.com/saudi-arabia/empowering-women-kingdom-leads-way.
- Vivian Nereim، Donna Abu-Nasr (2015, Augest 10th). Saudi Women Are Joining the Workforce in Record Numbers. Bloomberg Business: http://www.bloomberg.com/news/articles/2015-08-10/saudi-women-are-joining-the-workforce-in-record-numbers.
- Ahmed Yosri (14 Dec, 2015). Saudi Arabia elects its first female politicians. ALJAZEERA: http://www.aljazeera.com/news/2015/12/women-win-seats-landmark-saudi-arabia-elections-151213054750832.html
- Amy Tao. (N.d N.d, N.d). Thoraya Obaid. britannica: http://www.britannica.com/biography/Thoraya-Obaid
- Barbara Crossette (January, 2013). A Brave New Role for Thoraya Obaid, Former UNFPA Chief. passblue: http://passblue.com/2013/01/22/for-thoraya-obaid-former-unfpa-chief-a-brave-new-role/.

- Bashraheel, L. (2010). Al-Jeraisy: Role model for businesswomen. arabnews: http://www.arabnews.com/node/347752
- SPA (2011). AlRiyadh: http://www.alriyadh.com/626603.
- Diamond, W. (2015). Princess Reema Raises Middle East Breast Cancer Awareness With 10KSA Event. uffingtonpost: http://www.huffingtonpost.com/wendy-diamond/princess-reema-raises-middle-east-breast-cancer-awareness-with-10ksa-event_b_7968352.html.

A Global Problem with Poor Prognosis: Antibiotic Abuse

Danya Mohamed Bahanshal
Dental student, Al-Farabi Colleges, Saudi Arabia

ABSTRACT

This paper examines the level of knowledge the public have on antibiotics. No one can doubt the importance of antibiotic therapy and its major role in the treatment of bacterial infection, but are we abusing it? A major outbreak has recently occurred in many places around the world if not all, in the number of bacteria that are resistant to many antibiotics present. This outbreak has been attributed to the lack of proper education in the use of antibiotics, incorrect prescription and lack of regulation all leading to overuse. It is concluded from the present study that one of the major solutions is community awareness regarding antibiotics and its resistance to control this global problem.

Keywords: Antibiotic, Antibiotic Abuse, Antimicrobial resistance, penicillin.

1. INTRODUCTION

The development of antibiotics was a turning point in the world's life. It gave people hope and helped to extend their expected life spans. Penicillin was discovered in 1928 by Fleming and was used successfully by soldiers in World War II; shortly later on resistant bacteria appeared which resulted in a threat. Fleming then gave a warning in 1946 'the public will demand [the drug and] … then will begin an era … of abuses. The microbes are educated to resist penicillin and a host of penicillin-fast organisms is bred out which can be passed to other individuals and perhaps from there to others until they reach someone who gets a septicemia or a pneumonia which penicillin cannot save. In such a case the thoughtless person playing with penicillin treatment is morally responsible for the death of the man who finally succumbs to infection with the penicillin-resistant organism. I hope the evil can be averted' (New York Times 1945). This warning did not stop antibiotic abuse, antibiotics are overprescribed worldwide.
This resistance led to the development of new beta-lactam antibiotics. Unfortunately in the same decade methicillin-resistant Staphylococcus aureus (MRSA) was identified. This series of invention of new antibiotics followed by bacterial resistance continued until almost all antibiotics developed were resistant.
Figure 1. The Development of Resistant Bacteria, AB: Antibiotics
Resistance develops either by horizontal gene transfer (HGT) which transfers antibiotic resistance among different species of bacteria, or spontaneously through mutations. Once this happens these resistant bacteria increase in number as a result of natural selection.
To prescribe antibiotics, several factors should be considered such as the pathogens, patients and drugs. Unfortunately, this is not the case with most clinicians in which they tend to prescribe antibiotics without regulations or certification. On the other hand, specialists in oncology can only prescribe anticancer drugs.

This research aims to evaluate the level of knowledge the public have on antibiotics and raise awareness that is highly needed.

2. RESEARCH METHODOLOGY

This paper consists of primary and secondary data. Primary data was gathered by a survey questionnaire, which aims to evaluate public knowledge on antibiotic use in Saudi Arabia. This survey is composed of 10 key questions which are expected to measure if the

Figure 1. The Development of Resistant Bacteria, AB: Antibiotics

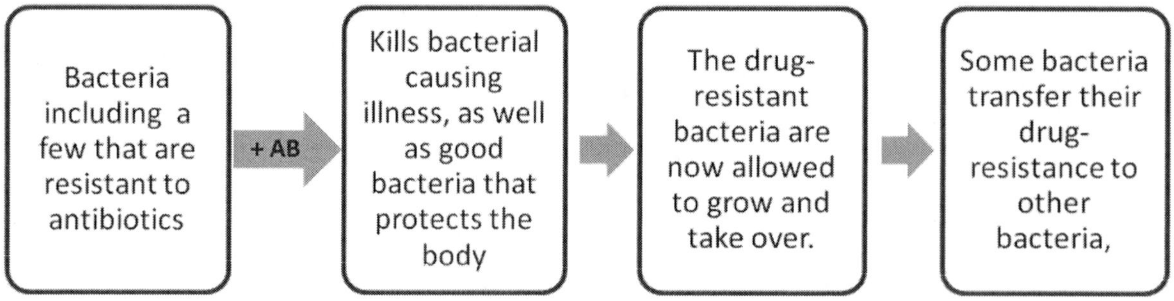

participant is aware of uses of antibiotics or not, and to identify whether they are used properly or not. The questioner ends by asking if the participants had a background on antibiotic resistance or not. It was sent to more than 300 participants, however; only 259 participants responded the questionnaire. The participants who answered the questionnaire are of different ages and education levels, which represent the public.

Secondary data was collected from PubMed database, academic journals and other medical websites.

3. FINDINGS OF SURVEY AND INTERPRETATION

3.1 Survey Results

The survey results shows, most of the respondents' age ranked between 18-24 with 34.75%, followed by 22.01% between 35-44 and 20.46% between 25-34.Around 22.78% age ranked between 45-74. From these respondents, 91.51% were female, while 8.49% were male. Most of the survey respondents' graduated from college with 55.98%, followed by 30.12% high school graduates, 13.90% healthcare-related education. When participants were asked how frequent they have taken antibiotics in a year? 45.95% indicate 1-2 times while, 21.62% took it 3-4 times, 16.22% more than 5 times and only 16.22% are those who haven't taken antibiotics. From the antibiotics taken 46.33% were prescribed by a dentist whether it was less than half or more than half of the antibiotics taken, while 53.67% of the antibiotics taken were not prescribed by a dentist.

Furthermore 54.83% of the respondents knew that antibiotics were used for bacterial infection while 36.68% thought it was either to treat colds faster or reduce pain.7.34% were not sure of why they took antibiotics and finally 1.16% took it just in case. Then participants were asked about whether they have taken antibiotics without a prescription? 49.03% of the respondent reported that they have sometimes taken antibiotics without a prescription while 41.70% never and finally 9.27% always take antibiotics without a prescription.

Later on participants were asked when they would end the prescribed antibiotic. 71.81% as prescribed, followed by 23.55% if feeling well after half the treatment and 4.63% are not sure when to end their treatment.

Moreover, to test participant awareness about the level of danger associated with antibiotics they were asked how dangerous can the side effects of antibiotics be. 63.32% responded that it may have a few, but nothing dangerous while 30.12% it may have serious side effect and 6.56% thought it was safe without side effects. Finally participants were asked whether they heard of bacterial resistance. 68.34% answered yes followed by 20.08% did not hear of it and finally 11.58% were not sure.

Figure 2. Descriptive Statistics

Gender	
Female	91.51%
Male	8.49%
Age	
18-24	34.75%
25-34	20.46%
35-44	22.01%
45-54	16.22%
55-64	5.41%
65-75	1.16%
Highest level of education completed	
Graduated from high school	30.12%
Graduated from college	55.98%
Healthcare-related education	13.90%

Figure 3. Survey Results

How frequent have you taken antibiotic in a year?	
1-2	45.95%
3-4	21.62%
More than 5	16.22%
0	16.22%
From the antibiotics taken, how much was prescribed by a dentist?	
Less than half	26.64%
Half or more	19.69%
None	53.67%
What is an antibiotic used for?	
Reduce pain	23.94%
Colds are cured faster by antibiotics	12.74%
For bacterial infection	54.83%
Just in case	1.16%
I am not sure	7.34%
Have you ever taken antibiotics without prescription?	
Always	9.27%
Never	41.70%
Sometimes	49.03%
How dangerous can the side effects of antibiotics be?	

It's safe with no side effects	6.56%
It may have a few, but nothing dangerous	63.32%
Serious side effects	30.12%
When do you end your prescribed antibiotic course?	
If you feel well after half the treatment	23.55%
As prescribed	71.81%
Not sure	4.63%
Have you heard of bacterial resistance to antibiotics	
Yes	68.34%
No	20.08%
I am not sure	11.58%

3.2 Analysis and Interpretation of Survey Findings

The study examines the general knowledge and behavior of the public towards antibiotic. The fact that 45.17% of the participants were either not sure or thought antibiotics were used for reasons other than bacterial infection shows the lack of knowledge of the exact use of antibiotics in almost half the participants. In many countries antibiotics are an over the counter drug and can be taken without a prescription or bought online results in overuse, this is highlighted by the fact that 49.03% of the participants sometimes take antibiotics without prescription and 9.27% of participants always take antibiotics without prescription.

63.32% thought that antibiotics have a few side effects but nothing important is an alarming result highlighting the immediate need for public education on the crisis we are passing through .As well as 20.08% of the participants have not heard of bacterial resistance at all. From the antibiotics taken 46.33% were prescribed by a dentist whether it was less than half or more than half of the antibiotics taken support the point that Dentist plays a major role in antibiotic resistance.

Over all, it can be concluded from the survey responses that we have a defect in the proper understanding of the function, uses, and the danger of antibiotics.

4. CAUSES OF OVER USE OF ANTIBIOTICS

There are several reasons for over use of antibiotics. C. Lee Ventola, MS (2015) stated that every year it has been accounted that the number of prescribed courses of antibiotics exceeds the population and thus counting for more than one treatment per person. The fact that in many countries antibiotics are an over the counter drug and can be taken without prescription results in the lack of regulation making antibiotics easily accessible, plentiful and cheap all leading to overuse. In countries were antibiotics are regulated the presence of antibiotic online makes them obtainable.
Another cause of abuse of antibiotics is its incorrect prescription from the beginning, which also results in resistant bacteria. For instance, according to C. Lee Ventola, MS (2015), Studies have shown that prescription regime is incorrect in 30% -50% of the cases. In intensive care units (ICUs) 30% to 60% of the antibiotics being prescribed are unnecessary, inappropriate or suboptimal. Moreover, The U.S Department of Health and Human Services (2015) noted that 50% of all the antibiotics prescribed for people are not needed or not effective as prescribed. Making antibiotics the most common prescribed drugs used in human medicine. According to Bartlett JG, Gilbert DN and Spellberg B (2015) a surveys show that 40% -75% of children and adults who are affected with viral respiratory tract infections take antibacterial agents as a treatment.

Incorrect Diagnosis and Diagnostic Tools leads to inappropriate antibiotic use when unsure of the exact diagnosis or till the diagnosis is determined, both results in the use of broad spectrum or multiple antibiotics in hope that one will be useful in treatment against the unidentified pathogen. This results in great harm to the patient and results in the development of antibiotic resistance. According to C. Lee Ventola, MS (2015) the development of a quick and accurate diagnostic tools such as the real-time multiplex polymerase chain reaction (PCR) and matrix-assisted, laser desorption/ionization, time-of-flight mass spectrometry could reduce this unnecessary use of antibiotics.

Dental general practitioners play a major role in this global problem. The British Society for Antimicrobial Chemotherapy (2004) published a review in the journal of Antimicrobial Chemotherapy, which concluded that dental practitioners cause a significant factor for the development of drug-resistant bacterial strains. A study by Cope, A., Wood, F., Francis, N., and Chestnutt, I. (2015) that was carried out reported that dental GP's regularly manage advanced dental problems of patients with a low financial income. The reason for this is sympathy with patients in dental pain that cannot afford dental consultation. In such cases GP's conceder antibiotics is an acceptable first-line treatment for these cases. GP's have realized that patients often attend expecting to receive antibiotics, and once prescribed a patient feels that a GP has given him/her their immediate needs.

5. CONSEQUENCES

With time studies it has shown a decrease in the development of new antibiotics due to economic and regulatory barriers. Bartlett JG, Gilbert DN and Spellberg B (2015) noted that from 18 large pharmaceutical companies, 15 have left the antibiotic field. For 4 decades there have been no antibiotics for gram-negative bacilli. Since 1998 only 2 drugs with a new microbial target have been made. This deficiency in production is associated with increase in resistance putting us in danger. The World Health Organization has stated that antimicrobial resistance is one of the three greatest threats to human health. Figure 4 illustrates a steadily decrease in the antibiotics developed by the pharmaceutical industries.

Figure 4. New Antibacterial Drugs Approved by the FDA

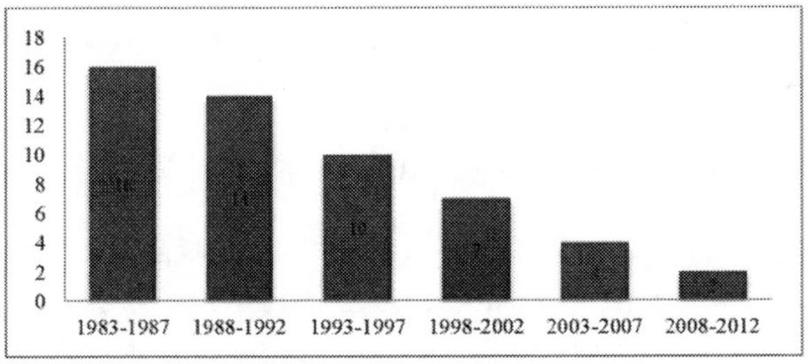

Physicians tend to prescribe old antibiotics and leave the newer agents on hold for a more advanced condition; this results in a decrease in production of new antibiotics. This decrease in use of new antibiotics results in a decreases in the return of investment in this costly procedure. Even when finance is not a concern, obtaining regulatory approval is often an obstacle faced by the antibiotic inventor. When options available for treatment are limited or are unavailable due to resistance, this puts pressure on the use of antibiotics that are toxic, more expensive and less effective.

Developing bacterial resistance is a very serious condition. Patients affected with resistant infections are at higher risk of staying longer in hospitals, death, and long-term disability. The American Association of Endodontists (2012) stated that the united states spend approximately $4.5 billion per year on hospital-acquired infections. Nosocomial infections are costly, increases morbidity, and results in longer hospitalization. Infected patients may develop antimicrobial-resistant organisms that may be transferred to hospital staff, hospitalized patients and later on to the community. C. Lee Ventola, MS (2015) stated that 'MRSA kills more Americans each year than HIV/AIDS, Parkinson's disease, emphysema, and homicide combined'. Other then the development of resistance to antibiotics, antibiotics have adverse drug events. According to U.S Department of Health and Human Services (2015) antibiotic resistance is responsible for one out of five cases in the emergency department.

5.1 Adverse Drug Events

5.1.1 allergic reactions

U.S Department of Health and Human Services (2015) stated that visitors of the emergency department for antibiotic-related adverse drug events are mostly due to allergic reactions, making up almost four out of five cases. Symptoms may vary from mild rashes and itching to serious blistering skin reactions, swelling of the face and throat as well as breathing problems.

5.1.2 C. difficile

Use of antibiotics results in the death of good bacteria for several months, which protects against infections. This may result the patient getting infected with C. difficile that may result in diarrhea. U.S Department of Health and Human Services (2015) notes that this results in the death of at least 14,000 Americans a year.

5.1.3 drug interactions and side effects

When antibiotics are taken with other drugs, they can interact with the other drug the patient is taking and may make it less effective or worsen the side effects of the antibiotics or the other drug. Some of the side effects of antibiotics can be nausea, diarrhea, stomach pain and dehydration.

6. SOLUTIONS

As noted by U.S Department of health and human services (2015) 'With over 2 million infections – and 23,000 deaths – caused by antibiotic resistant bacteria each year in the US, everyone agrees there is an urgent need to find new, effective antibiotic therapies'. We all have to step up to find solutions to not only save lives but also enormous amounts of money for this crisis. Since antibiotic resistance is a natural process it can't be stopped but can be slowed down, so there is always a need for new antibiotics.

6.1. Prevention

A very logic solution is to try to avoid infections in the first place, which will reduce the amount of antibiotics needed, resulting in a reduced chance in developing resistance during therapy. There are several methods to reduce infections. This includes immunization, safe food production and hand washing.

6.2 Education

6.2.1 education of healthcare professionals

Effort has been put in the education of physicians in hospitals. Chang-Ro Lee, Jung Hun Lee, Lin-Woo Kang, Byeong Chul Jeong, and Sang Hee Lee (2015) noted that the antimicrobial stewardship program has multiple strategies with education being one of them. Since hospitals are complex institutions, methods such as one-on-one educational sessions given to physicians may be more effective. Moreover, postgraduate educational courses, training antibiotic prescribers in the community is also of great importance.

6.2.2 education of undergraduate students

Little attention has been given to the education of undergraduate students in the medical field. Several studies have shown a great defect in their knowledge of when and how to use antibiotics. Students with different sources have different level of education on this topic. The development of standard formal curricula about antimicrobial use and resistance is required.

6.2.3 education of the public

When conducting the survey, it showed the presence of a great defect in public knowledge about antibiotics. Only 54.83% knew that antibiotics are used for bacterial resistance while 23.94% of the respondents agreed that antibiotics could be used to relief pain. When participants were asked on the side effects of antibiotics only 30.12% agreed that it has serious side effects while 63.32% agreed that antibiotics have few side effects but nothing dangerous. These results from 259 participants in Saudi Arabia have shown that effective public education has to be provided. European Antibiotic Awareness Day and Get Smart: Know When Antibiotics Work in the United States are campaigns that have been conducted, so why not start one in the Kingdom of Saudi Arabia.

6.3 Stewardship Program

Antibiotics are a limited source using it more now makes it less likely to be effective in the future. That is why following a strict guideline in the use of antibiotics (antibiotic stewardship program) is mandatory. Stewardship is a commitment to use antibiotics properly as needed with correct prescription. This will ensure that each patient will be given the maximum benefits of antibiotics and avoid the unnecessary consequences of antibiotic abuse. Several studies have shown great benefits and reduction in total antibiotic consumption, duration, and inappropriate use with antibiotic stewardship program.

6.4 Tracking

U.S Department of Health and Human Services (2015) stated that infections, antibiotic use, and resistance are reported by healthcare facilities and sent to the Centers for Disease Control and Prevention (CDC) which help in Constant tracking and collection of data on antibiotic-resistance, their causes and possibilities of particular risk factors. This helps experts develop strategies to prevent infections and their resistance from spreading.

6.5 New Regulatory Pathways

Antibiotics have a negative net present value (NPV); it makes them an uneconomically wise investment. This is due to many factors including that they are given in short courses, generally have low prices per course compared with other drugs that are often given for a decade. The FDA has proposed new regulatory pathways to regain motivation and facilitate the production of antibiotics making them less expensive take less time and address a critical need.

6.6 (GAIN) Act

The Generating Antibiotic Incentives Now (GAIN) Act is another important act that helped improve the NPV, which was signed into law as part of the FDA Safety and Innovation Act. The antibiotic being invented can enjoy extra 5 years guaranteed market exclusivity with no competitors such as another version of the same drug, as the FDA will reject it even if the drugs patent life expired (Seaton, D. ,2013). Even though this is a strong first step a lot still needs to be done.

6.7 Delayed Prescription

'Delayed prescription' is a Cochrane review (2005) which means that any antibiotic prescribed is given a few days later to give time to observe weather symptoms improve or not. Such a method helps satisfy both patient and prevent unnecessary use of antibiotics.

7. CONCLUSION

Antibiotic resistance can easily cross large distances, making this issue an international concern. Good news is that this can be reversed so there is still hope. The World Health Organization (WHO) emphasized both public and professionals play a role in the control of antibiotic resistance. For those that have established their knowledge and behaviors about antibiotic use, making a change is very difficult. So that is why constant education especially for children at a young age is more effective.

From the time antibiotics were discovered, it has made treatment of deadly diseases possible. So let's use it properly so that this 'double ended sword' does not act against us.

REFERENCES

- Andre, M., Vernby, A., Berg, J., and Lundborg, C. S. (2010). A survey of public knowledge and awareness related to antibiotic use and resistance in Sweden. *Oxford journals , 65 (6),* 1292-1296.
- Bartlett, J. G., Gilbert, D. N., and Spellberg, B. (2013). Seven Ways to Preserve the Miracle of Antibiotics. *Oxford journals , 56 (10),* 1445-1450.
- Lee, C. and Ventola, M. (2015, April). The Antibiotic Resistance Crisis Part 1: Causes and Threats. MediMedia USA, Inc. , 277–283.
- Lee, C. and Ventola, M. (2015, May). The Antibiotic Resistance Crisis Part 2: Management Strategies and New Agents. MediMedia USA, Inc., 344–352.
- Chang-Ro Lee, J. H.-W. (2015). Educational Effectiveness, Target, and Content for Prudent Antibiotic Use. BioMed Research International , 1-13.
- Cope, A.L., Wood, F., Francis, N.A., and Chestnutt, I.G. (2015). General practitioners' attitudes towards the management of dental conditions and use of antibiotics in these consultations: a qualitative study. *BMJ Open , 5 (10).*
- DaVolterra (2015). Protecting Intestinal Microbiota From Antibiotic-Induced Disruption. Retrieved from DaVolterra: http://www.davolterra.com/content/protecting-intestinal-microbiota-antibiotic-induced-disruption.
- Endodontists, A. A. (2012). Use and Abuse of Antibiotics. Endodontics Colleagues for Excellence.
- Krans, B. (2015). Few New Drugs: Why the Antibiotic Pipeline Is Running Dry . Retrieved from Healthline: http://www.healthline.com/health/antibiotics/why-pipeline-running-dry
- Luyt, C. E., Brechot, N., Trouillet, J. L., & Chastre, J. (2014). Antibiotic stewardship in the intensive care unit. critical care , 18 (5).
- Najla Saeed Dar-Odeh, O. A.-H.-O. (2010). Antibiotic prescribing practices by dentists: a review. Dovepress , 6, 301−306.
- Seaton, D. (2013, November 10). GAIN Act: A Great First Step. Retrieved from BIOtechNOW: http://www.biotech-now.org/health/2013/10/gain-act-a-great-first-step.
- Services U. D. (2013). ANTIBIOTIC RESISTANCE THREATS in the United States, 2013. Centers for Disease Control and Prevention .
- Sweeney, L. C., Dave, J., Chambers, P. A., and Heritage, J. (2004). Antibiotic resistance in general dental practice—a cause for concern? Oxford Journals , 53 (4), 567-576.
- The 10 x '20 Initiative: Pursuing a Global Commitment to Develop 10 New Antibacterial Drugs by 2020. (2010). Oxford journals , 1081-1083.

A Name Starts: A Step for Development in Africa

Thabang Morena Letlabika
Local Economic Development (LED) Intern, Joe Gqabi District Municipality, South Africa

1. INTRODUCTION

Leadership? We can have so many definitions of what makes a great leader, prime minister, king or class captain. I have seen the great potential that the youth around me inherits, yet we as youth are never mentored to be great at the end of the day.
I am an African whose forefathers have been subjected to colonisation and after decades of striving for new possibilities they accomplished it. Inevitably after generations of hard work we are still faced by the many socio economic issues that are common in my community to Africa, Asia, Latin America and the rest of the world.

The National Youth Policy of South Africa defines youth development as: an intentional comprehensive approach that provides space, opportunities and support for young people to maximise their individual and collective creative energies for personal development as well as development in a broader society of which they are an integral part.

I have valued and cherish the dreams of those that surround me, yet as educated as I am, I am still faced by the problems of many in Africa and internationally, unemployment. It is so that I am indebted in changing the lives. *'I am an African. I owe my being to the hills, the valleys, the glades, the rivers, the deserts, the trees, the flowers ,the seas and the ever-changing seasons that define the face of our native land'*, former President Thabo Mbeki. It is so that I am compelled to this being and the forever existing socio economic issues that we as Africa and the rest of the world are faced.

2. RATIONALE FOR A NAME STARTS

'AIDS' (Africa Is Dying Slowly). At the outset of great leadership and potential this document promotes the greatness that rests in the youth of today. We are a discouraged youth faced by undefined parameters of race, age, gender, social class and geographic location. We are a youth that needs to be mentored and harnessed for greatness.

A 'name starts' envisions the great power of networking and communication through various intuitions to ultimately changing lives through influence. It is about real life situations I have encountered around my community, country and how these common problems are common around the world. It sets to envision and path a way set by National Development Plan of South Africa.

3. A NAME STARTS

I have seen the hardships of poverty and inequality that surrounds me. A name starts when one sees the need to make a difference, not for him nor peers or those that surround him. It has demented generations that services to those of the poorest poor have no access to fully access them. It is not only today that I thought I should write this. What makes a name?

Over the past two decades slowly change has begun. Post 1994, South Africa, this nation was granted freedom change began. Inequalities still exist yet it seems inevitable to change this. Unemployment levels increased over the second quarter of last year, 2014 increased to 25.5 %. This has been the highest peak for unemployment in this country from a 22 per usual-25.5 per cent peak. I am from a province where unemployment has reached a maximum peak of 27.2 %, in 2009, this high unemployment rate is influenced by the fact that it is mostly rural. (*Development Trends in Joe Gqabi District Municipality 2010*).

This is yet another problem in one country that is common in the rest of the world. A name should start when one sees the need to turnaround this situation and use what is already given.

During the course of April 2015, I met an old woman in passing, not noticing that she was blind. Her home was wired all over with cotton wool, it puzzled me as to what was happening, only after a while it got to me, this was her way around the yard in doing her basic necessity. Through this I had seen that that old woman had lived alone most of the time, in instances such as these it reflects on the many problems that we face on social security and poverty. This situation gives us a partial overview of the socio economic problems that we face. It saddened me that through so many struggles that yet again so many people's lives have not changed.

My aim through this document is to reflect on real life situations around me, with the hope that through communication and networking rural lives are changed. It sets to promote a new way of thinking through valuing skills and the great potential that rest in the youth. It is a way forward to relief those still stuck by cultural norms in progression. It shares a desired common goal by community activist in lives and what needs to be done. This name could start.

3.1 A Lost Generation

During the course of 2015 30th July a summit was held on the Transforming African Universities. Firstly what are universities and what are its core their intents? Tradition universities started as research facilities and over the years they have diversified to be more than that. Universities today stand as ground to shape society and develop flexible or future identities (Cross et al. 1999, Kerr 1991, Castells 2001 & 2009). This is acquired through education and skills development programs. This summit heightened the important role that the state and its universities have in acquiring shared common goals.

I am from a lost generation where only matric is still valued as the highest form education. By the age of 18 on average a young girls is negotiated for lobolla (hand in marriage) and a young man sent off to Johannesburg in search of a job or to serve as a provider of the household. Sometimes the man never finds a job. However if such ever happens these individuals are solely dependent on government grants and subsidies. I have had the pleasure to meet those privileged of the many opportunities offered to them. In most instances some of us privileged of this power (education) drop out during the course of undergrad degrees. We are again a lost generation. Most of those I know have ventured out universities in search for entrepreneurial ventures, after this we struggle getting funds or even to start, and the end result being frustration and also contributing to this epidemic of unemployment and its effect on communities.

From the above scenarios there are two kinds of people, those informed and not informed. So what should exactly happen to address these two kinds of people? Are we a lost hope or present opportunity?

4. A FOUNDATION FOR CHANGE

Sterkspruit, also forms part the Joe Gqabi District Municipality in the Eastern Cape, South Africa. It is surrounded by the majestic Drakensburg Mountains, diverse cultures and tribes. It is home for more than 325 000 people. Over the past years this population estimate has decreased from 350 000 est. to 324 840 est., this is predominantly from migration reasons and HIV/other diseases. This district consists of a high majority of unskilled labour and a shortage of skilled labours (*Development Trends in Joe Gqabi District Municipality 2011, Ed.). It is a rural area like most in the country still stagnan*t in developing.

This province is home to a number of well-known international leaders such as Nelson Rhlolihlahla Mandela, Steve Biko and former President Thabo Mbeki, one Africa's most active leaders in the African Union. Currently Sterkspruit it consists of three main professions teachers, policing and nurses.

What defines change? I am rooted with strong cultural beliefs and in this regard it is within my nature to fear to contradict these beliefs. I have seen people and generations stagnant on the same measures of living, similar profession being carried out through their generations. It has been a norm of thinking that has not changed.

The world cannot be changed without changing our thinking϶, Albert Einstein. What defines change? A new way of thinking. I have had the opportunity to connect with those mostly privileged, yet again I have grown with those stuck by cultural norms and beliefs. In all that has come I have not changed from my roots yet I have seen a foundation that could change the norms of those.

I am surrounded by few opportunities of exploitation, but it takes someone to truly be in a situation to understand where they from and where they want to be. I am from a hometown where villages are named by surnames of the most prevalent in that area. It is a hometown with that still has a vast amount of people living in traditional dwellings, 31.7 % and 1.7%

informal and unspecified 1.7%. Where livestock still measures the wealthiest family or individual. See below. Figure 2, AG chrome vibe.

Change needs to happen on a small scale before it can proceed to a larger audience of society (a bottom up approach). It is therefore that rural areas/towns need to be developed. Sterkspruit consists with high majority of young and elderly people. Mainly for this reason, in search for better standards of living people move where more opportunities exist, the results of this is either young adults leave behind their children or elderly people. Sometimes none of them ever return, leaving those behind stranded and resorting to other means of making a living. These members of society are the once that a most exposed and vulnerable. Ideally I am to reflect on real life situations that apply around Sterkspruit and with the hope that they apply to other parts of the country and the world.

We are rural town that inherits more than 80 rural dwellings and a majority still dependant of livestock farming and government grants/subsidies. A minority has had the privilege to make it to matric and finish high school whilst only a minority make it to tertiary education level. Most of these families only take their young once up to matric and it is highly valued by their families. These are some of the reason why this great nation falls behind in education in comparison to other states of the world. In recent ratings we it is rated no.148 by the World Economic Forum,2014. In analysing this ,one should be clear that not much is done to change this, this is why in this passage I should value skills, solely for the reason that a vast majority of our youth never proceeds to tertiary education. By skills I refer to basic skills that one can grasp easily or can be taught.

The National Development Plan of South Africa is a growth policy, an act that embodies our current governments; it aims to achieve a growth rate of 5 per cent by 2030. It defines a desired destination, only if all sectors play their role in attaining these goals .It aims to redress poverty and inequality. It sets new form of leadership that is transparent and renders accountability. This is a great policy with a clear vision, yet it has not taken its full effect.

During these crucial days of being unemployed I valued the efforts of national/provincial governments in fighting the battle against unemployment through community driven initiatives, in so I also analysed the people that worked in these projects. EPWP (Expanded Public Work Programs) and CWP (Community Work Programs) are government initiatives, part of the NDP, set to address the problems that exist in our communities. It set to account 55% woman working, 3% with disabilities which include both men and woman, 30% youth, 10-12% consideration. It indeed a good policy, but at the end of the day many of the individuals working in these programs are adults above 35 years. How do we fully reach a target youth of 30 %? Given that our youth is selective/ discouraged.

It is imperative that as youth we notice the huge role that we play in shaping our communities thus ultimately our country, unemployment is therefore not only governments' sole responsibility but each and everyone's commitment. A way forward would be in shaping new strategies and innovative ways to create a foundation of change. A need would be a 'youth organisations' in opening doors and setting out channels of networking to reach suitable stakeholders in government and the private sector.

This body would discuss relevant issues affecting their towns/communities/states. It would network channels on behalf of the community to stakeholders. It would be subsidiary of well-known non-governmental organisations such as the G200 Association or a combination of the different bodies within society. It should reflect on what is happening within communities/towns/cities and sets best interest. In the long run if possible it would centralise data for existing and potential graduates, it encourage and centralise university applications for matriculates in assisting them in application processes.it would endorse information sharing sessions to the public in applying for funds/projects for the development of communities. It is not a political organisation. (Figure3).

4.1 The Organisation

Name ....

Founder ....

Vision and Mission Statement ....

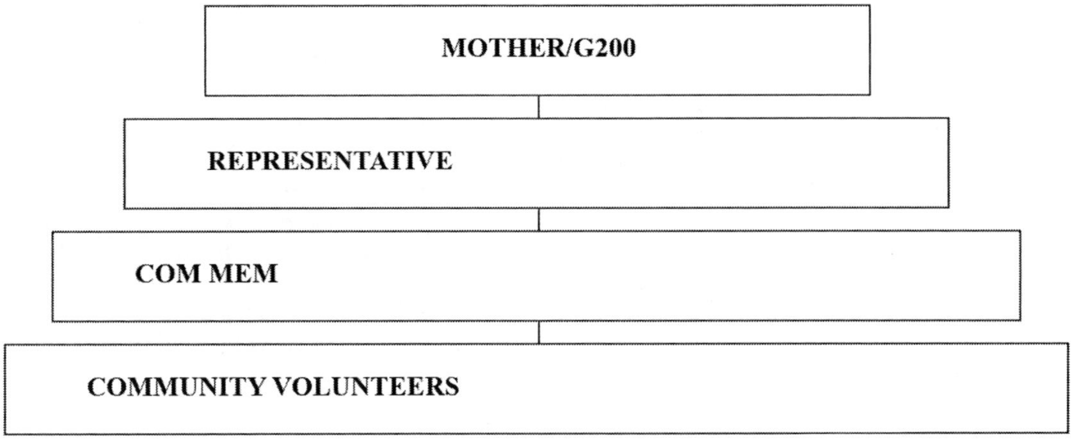

5. A WAY FORWARD

5.1 Change Attitudes of People through Effective Leadership

It only natural that a child mimics what an adult does. It is common amongst Africans that our leaders often overstay their terms of power. This has had detrimental effects on the communities themselves and neighbouring states. Earlier through this article I expressed the imperativeness of a youth that needs to be mentored and harnessed for effectiveness.

Mentorship – defined as a personal developmental relationship in which a more experienced or more knowledgeable person helps to guide a less experienced or less knowledgeable person.

Harness - to control and use force or strength to produce power to achieve something.

I am surrounded by a vast amount of potential (knowledge/power) yet at the end of the day so many of us never discover our capabilities. 'Education is the most power weapon you can use to change the world', Nelson Mandela. We are none the less a misinformed youth on what happens around the world, a key to start changing lives is through information. An art of giving back and volunteerism should also be rooted in our communities, as against the culture of people making it through their lives and never look back. These people should be awarded.

I believe as a leader you need to change the way people perceive things and mostly enlighten them of a way forward. As a leader your people should come to realisation of their current situation and accept it (what surrounds them), this is vital in the sense that people can start to see a need within their communities and grasp on those opportunities. This can only be done if information is viably shared.

5.2 Increase Investments

I think this is the most important aspect for development, acquisition of capital. Earlier in this article I described Sterkspruit as a rural town that has a vast majority of elderly people plus 60 years. These individuals are mostly pensioners from three professions, nursing, teaching and policing. Once government pays out a lump sum to these two individuals, people either renovate their homes or buy cars and never prepare for a long term solution. We are a town that lacks knowledge on financial planning and investment management. These groups of professionals can contribute economically in these areas together with their local municipalities.

These people can venture in: Pooling Joint funds for investment purposes in establishing property, invite possible companies to render their property to sell goods or render services. These people could share life time benefits in rent income or acquiring shares/ proportions in these businesses. Wealth can therefore be created through these channels. Spill overs avoided

(Money can be circulated countless times in one area thus creating wealth).

5.3 Capturing of Community Projects/ Youth Driven Initiatives

I outlined a lost generation, I am from rural village were a young person can be a herdsman throughout his life. Such people a deserving of community projects. Astonishingly the informal sector consists with a vast amount of potential individuals that can run businesses effectively. A feasibility study should be conducted to identify such individuals.

I believe that talent search programs should not only be focused in major cities rather across rural towns and areas. Through this talent is harnessed and developed.

5.4 Universities /Private/Government Institutions Canvas.

Graduating and not knowing where employment lies is stressful. I have been unemployed for plus 8 months, it has been hard that starting up fresh is not as easy as it should.

Out seeing this problem made me realise that this as a problem that can be dealt with on graduation day. Private/Government sectors should effectively canvas themselves for internships/jobs on graduation day. Youth graduate fill in forms with the hope that they will be contacted. I think this should be an agreement between government/private sector and universities. Private sector companies should also put emphasis on rural canvasing. Municipalities could make names for themselves.

Universities should start going to high schools and rural towns to motivate young people of opportunities they have for them and possible benefits of studying their institutions. Such programs should exist every second quarter just before exams to remind youngsters, this could motivate them. Alternatively governments in local municipalities can centralise a data specially dedicated for graduate with that district municipality.

5.5 Take Jobs to the People.

Ironically black people still associate a white man with a job opportunity. I was saddened that one of my peers white man in passing and asked him if he wanted workers. It is still an African norm. During the course of my article I mentioned people that left their homes in search for job opportunities in major cities around the country, resultant effect is it leaves young once depressed and having to provide for themselves. During May 2015, vacancies opened up; sadly most of us did not have transport money to go to town. I believe that similarly how my friend associated a white man with a job, we should have something from private or public sector that we can associate our chances of employment with. This builds confidence to our people and stakeholders.

5. CONCLUSION

As population growth increases there can never be sufficient jobs available. This is a partial situation in a rural town that can be aligned with the problems that societies/communities around the world face. It is therefore an ideal for effective leadership to bring about change within our communities. It is a matter of advancing (rural lives) to first world economies and lastly, the first steps to development in any rural society are acquiring the basics .Therefore education and information is power and key to development in any rural community.

REFERENCES

- African Union/NEPAD (2005). Renewal of Higher Education in Africa. Report of AU/NEPAD Workshop, 27-28 October 2005, Johannesburg.
- World Bank (2007). World development report 2007: Development and the next generation. Washington, D.C: World Bank. Retrieved from: http://www-wds.worldbank.org.
- Dr. Thriven. Higher Education and Social Transformation, South Africa Case Study: Reddy (University of Cape Town).
- Development Trends in Joe Gqabi Municipality: Socio Economic Profile Report (2011).
- Statistics South Africa (2014). Labour Force Survey, Quarter 2, P0211. Pretoria.
- Cloete (1997). Knowledge, Identity and Curriculum, Transformation Africa.
- World Economic Forum. Retrieved from: www.weforum.org/community/forum-young-leaders.
- Castells, M. (2009). Universities in Development, the Economy and Society. Transcript of a lecture given by Manuel Castells at the University of the Western Cape, August 2009.
- Kerr, C. (1991). The Great Transformation in Higher Education, 1960-1980. Albany: State University of New York Press.
- Remarks of the patron of Transforming African University (TMF), Thabo Mbeki Speech, Africa Summit: University of Johannesburg, 30 July, 2015.
- African National Congress. 'I am an African' - Thabo Mbeki's speech at the adoption of the Republic

of South Africa Constitution Bill. 8 May 1996. Cape Town. Retrieved from: www.anc.org

- Ratings: Economy. Retrieved from: www.fin24.com/Economy
- David, B. (1997), Clark (2004), Geiger (2004). Higher Education in the World 3.
- Robins, G. Retrieved from: www.cs.edu/~robins/quotes
- Levine, D.K. and Modica, S. Why Nations Fail: A review of Acemoglu and Robinsons.
- Piketty, T. (2014). Capital in The 21st Century. Paris School of Economics.

A Double-edged Sword: The Effects of Challenge Stressors on Job Performance

Yichun Kong
Student, School of Business, Nankai University, China

ABSTRACT

Despite acknowledging the effect of challenge stressors in organizations, researchers have rarely explored its ambidexterity. Previous findings suggest challenge stressors are positively related to job performance. However, existing researches demonstrated mixed results. While many scholars have found that challenge stressors are associated with beneficial effects, others have found it harmful. To resolve this contradiction, the study articulates a model that includes meaningfulness and exhaustion to examine how challenge stressors affect the job performance, and the likely performances follow in these two paths. When challenge stressors are within an appropriate range, meaningfulness motivates individual to work hard and make a better performance. However, when challenge stressors keep increasing and exceed an appropriate level, employees tend to achieve few outcomes because of exhaustion. This paper proposes an inverted U-shape relationship between challenge stressors and job performance and examines the moderating roles of competence in the relationships between challenge stressors and job performance. We conclude with a discussion of how the curvilinear indirect effect of challenge stressors works on job performance, as well as implications for future research and practice.

Keywords: Challenge stressors; Job performance; Meaningfulness; Exhaustion; Competence

1. INTRODUCTION

Stress in the workplace has been a significant area of research due to the impact on both organizations and employees (Barling, Kelloway & Frone 2005). The constantly increasing amount of stress in workplace that employees are experiencing has attracted widespread attention. According to Cavanaugh and colleagues (2000), relationships between stressors and job performance are different according to the type of stressors (e.g., challenge-hindrance stressors). Challenge stressors were defined as 'work-related demands or circumstances that, although potentially stressful, have associated potential gains for individuals' (Cavanaugh, Boswell, Roehling & Boudreau 2000, p.68). Challenge stressors include high workload, time pressure, job scope, and strong sense of responsibility, but these stressful demands have potential benefits to one's personal growth and career development. By contrast, hindrance stressors were defined as *'work-related demands or circumstances that tend to constrain or interfere with an individual's work achievement and that do not tend to be associated with potential gains for the individual.'* (Cavanaugh, Boswell, Roehling & Boudreau 2000, p.68). Hindrance stressors can be viewed as unnecessary obstacles to personal growth, because they produce role ambiguity, role conflict, organizational politics, job insecurity to workload (Cavanaugh, Boswell, Roehling & Boudreau 2000; Lepine, Podsakoff & Lepine 2005; Jackson & Schuler 1985). Both types of stressors have negative relationships with performance through strains (e.g., Boswell et al. 2004; Lee & Ashforth 1996; LePine et al. 2004; LePine et al. 2005; Podsakoff et al. 2007; Webster et al. 2011; Lin et al. 2014).

Previous findings suggest that challenge stressors are positively related to performance (e.g., LePine et al. 2004; LePine et al. 2005; Pearsall et al. 2009; Wallace et al. 2009). However, existing researchers have found conflicting results. Some empirical studies and meta-analysis have shown that challenge stressors have no significant correlation, even negatively, with job performance (Lin, Ma, Wang & Wang 2015; Liu, Liu, Mills & Fan 2013; Webster, Beehr & Christiansen 2010). The study proposes that there is a nonlinear relationship between challenge stressors and job performance.

According to the work of Lepine (e.g., Lepine, Podsakoff & Lepine 2005), challenge stressors are positively related to motivation. Through motivation (e.g., meaningfulness), challenge stressors could evoke a positive work attitude (Podsakoff, LePine & LePine 2007). Therefore, challenge stressors are positively related to job performance. However, when challenge stressors keep increasing and exceed an appropriate level (e.g., challenge stressors from medium to high level), it may lead to the mental exhaustion. Employees tend to achieve fewer outcomes under higher level of stressors. In such situations, challenge stressors are negatively related to job performance. Accordingly, the current research proposes an inverted U-shape relationship between challenge stressors and job performance.

We use the transactional model of stress from Lazarus & Folkman (1984) to illustrate the potentially nonlinear relationship between challenge stressors and performance. It is logical to take into account individual competence differences. Lazarus (Lazarus 1991; Lazarus & Folkman 1984) proposed that cognitive appraisal process including primary appraisal and secondary appraisal. At the primary appraisal, people appraise an event as challenging, hindering or irrelevant. The primary appraisal perceives the importance and relevance of environment to the person. At the secondary appraisal, people appraise the coping responses in order to reduce the negative effects of the event. When facing a moderate level of stress, individuals will cope positively and perform well. However, when faced with the increasing levels of workload which is beyond the competence, individual will suffer performance declines. Findings are consistent with the hypothesis that the nonlinear relationship between challenge stressors and performance variables is moderated by competence. For highly competent individuals, challenge stressors could act as an important support for potential gains to handle the high workload and boost job performance. In contrast, individuals with low competence might slack off at work when facing the excessive challenge stressors, which could lead to low job performance. This study posits that individual competence will weaken the inverted-U relationship between challenge stressors and job performance.

In sum, by empirically investigating the nonlinear relationship between challenge stressors and performance, as well as the moderating effect of competence, the study contributes to the literature in three ways. First, in the literature on stressors, scholars have tended to view challenge stressors as 'beneficial'. However, many organizational scholars have disagreed on whether its impact is helpful or harmful. Previous researches demonstrated mixed results and found that there are some situations where challenge stressors could exacerbate job performance. By identifying the inverted-U shape relationship between challenge stressors and job performance, the study extends the existing theory and provides a new insight to understand the impacts of challenge stressors. These findings extend stress literature by clarifying the boundary conditions of the challenge stressors-performance relationship, which highlights the importance of individual competence differences in the stress management area.

Second, as depicted in Figure 1, this study articulates a more nuanced model. By discussing the mediating role of meaningfulness and exhaustion on job performance, the study proposes that challenge stressors have a curvilinear relationship with meaningfulness (inverted-U shape) and exhaustion (J-shape). A more nuanced model of challenge stressors-performance process will help to understand why previous studies could not support this argument consistently.

Third, the study also contributes to both stressors and job performance literatures by considering the role of individual competence. By examining the moderating role of competence, the study proposes that competence moderates the curvilinear indirect effect of challenge stressors on job performance through meaningfulness and exhaustion. As such, our study offers a more nuanced picture of how challenge stressors affect the job performance.

Figure 1. Hypothesized Research Model

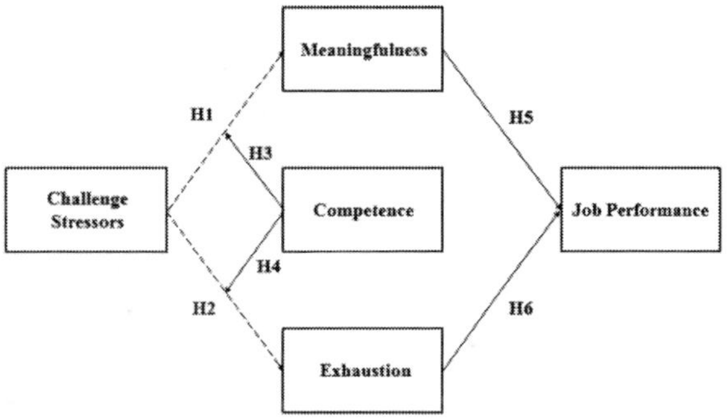

Note. H=Hypothesis. Dashed lines represent nonlinear effects (i.e., H1 and H2). H5 and H6 represent nonlinear indirect effects of challenge stressors on job performance.

2. THEORETICAL BACKGROUND AND HYPOTHESES

According to the transactional model of stress (Larzarus & Folkman 1984), the primary appraisal refers to the evaluation of individuals on the potential impact of stressors. When individuals perceive events are not likely to affect outcomes, events will be appraised as 'irrelevant'. When events are perceived as an opportunity for personal growth, development and well-being, they will be appraised as 'challenging'. When events are perceived as taxing to goal attainment, they will be appraised as 'hindering'. As such, when challenge stressors are within an appropriate range (i.e., from low to medium level), people are inclined to make positive evaluations and appraise these work events as challenging. The increasing challenge stressors could enhance the level of 'challenging' appraisal. When individuals believe that challenging work events could potentially promote their personal growth and career development, they would show a high understanding of the meaningfulness of their job.

On the other hand, however, when challenge stressors exceed an appropriate level (e.g., medium to high level), people tend to make negative evaluations and appraise these work events as hindering. When challenge stressors exceeding acceptable levels of stress, the evaluations of 'hindering' are intensified. With the hindering appraisal becoming higher and higher, people tend to believe that these work events would hinder rather than facilitate their personal growth and career development because they can hardly deal with such high burden of work events. In turn, meaningfulness might decrease.

Hypothesis 1. Challenge stressors have a curvilinear relationship (inverted-U shape) with meaningfulness.

Exhaustion indicates *'feelings of being emotionally overextended and depleted'* (Leiter et al. 1998, p.69) and is considered as *'the central quality of burnout and the most obvious manifestation of this complex syndrome'* (Maslach et al. 2001, p. 402). When challenge stressors are appropriate, there is no significant relationship between challenge stressors and exhaustion, because people tend to make positive appraisal and work hard. However, when challenge stressors are excessive or the investment of resources does not lead to performance gain, individuals are likely to appraise them as hindering. As challenge stressors keep increasing, individual' emotional resources become depleted, and they no longer identify enduring qualities as expected. In situations of high challenge stressors, the relationship between challenge stressors and exhaustion is positive.

Hypothesis 2. Challenge stressors have a curvilinear relationship (J-shape) with exhaustion.

According to the transactional model of stress (Larzarus & Folkman 1984), the secondary appraisal refers to identifying the potential appropriate coping response to the specific event. Individuals determine whether coping strategies and actions can be taken to reduce the detrimental impact. If stressors are perceived as challenging, methods of problem-solving such as preparation, allocating time for projects will be taken. In contrast, if the appraisal of stressors is hindering, it will elicit more forms of coping. Hindering appraisals *'can encourage withdrawal or defensive operations that turn the person inward or encourage hostile, aggressive behavior'* (Lazarus & Folkman 1984, p. 191).

Competence moderates the curvilinear relationship (inverted-U shape) between challenge stressors and meaningfulness. Highly competent people are inclined to evaluate events as challenging and tend to be confident to cope with it. In contrast, less competent people are more likely to appraise the stressors as hindering and feel difficult to cope with it. Therefore, when individual competence is high, it moderates the curvilinear relationship between challenge stressors and meaningfulness such that the inverted-U shape is mitigated. When a person has lower competence, the inverted-U shape will be magnified.

Hypothesis 3. Competence moderates the curvilinear relationship between challenge stressors and meaningfulness such that the inverted-U shape is mitigated when competence is high.

Competence moderates the J-shape link between challenge stressors and exhaustion. People who are highly competent may be more likely to select more effective coping strategies, because they feel excited and joyful during this period. Highly competent people tend to appraise the situation as more of challenging; thereby decreasing perceived stressors effect on exhaustion. Therefore the relationship (J-shape) between challenge stressors and exhaustion is weak. However, people with less competence are more likely to feel difficult to deal with the situation, and appraise events as hindering. Increasing perceived stressors effect on exhaustion strengthens the relationship (J-shape) between challenge stressors and exhaustion.

Hypothesis 4. Competence moderates the curvilinear relationship between challenge stressors and exhaustion such that the J-shape is mitigated when competence is high.

Previous research has clearly shown the linear relationship between meaningfulness and performance (e.g., Hunton & Price 1997; Richard & Richard 1977; Jon, Stephen & Susan1991). Varied levels of personal engagements are determined by perceptions of the meaningfulness (e.g., William 1990). Responding to high perceptions of the meaningfulness in their job, people put more resources into their work ,and make a better quality of outcomes. Also, drawing on previous theories (Jeannie & John 1983; Wright & Cropanzano 1998), the study acknowledges the linear relationship between exhaustion and job performance. People who experience high exhaustion will reduce the level of effort (i.e., time, energy), which makes it difficult to put available cognitive resources on the most relevant aspects of a work project. As a result, emotional exhaustion is associated with decreased productivity. Job performance in the face of challenge stressors is shaped by the appraisal individuals make of their work situation. Drawing on this logic, the study pays attention to the indirect effects of challenge stressors through meaningfulness and exhaustion to job performance.

For people who are highly competent, hypothesis 3 suggests that the positive effects of challenge stressors on meaningfulness will be enhanced and the negative effects will be mitigated. Given the positive influence of meaningfulness on job performance, the study argues that challenge stressors (from low to medium level) will make a positive but indirect effect on job performance. As high challenge stressors hinder the personal growth through meaningfulness and exhaustion, we expect the indirect effect will be insignificant at such levels. For people of less or lower competence, the preceding hypotheses indicate that challenge stressors enhance the negative effects on meaningfulness and exhaustion (Barrick et al. 2010). The relationship is insignificant at low levels while negative at intermediate to high levels of challenge stressors.

Building on this logic, the inverted-U shape effect of challenge stressors is expected to be transmitted through meaningfulness and exhaustion such that the indirect effect on team performance is insignificant at low levels of challenge stressors and negative at intermediate to high levels of challenge stressors. In sum, the mediated inverted-U shaped challenge stressors→meaningfulness and exhaustion→job performance relationship will be moderated by competence.

Hypothesis 5. Competence moderates the curvilinear indirect effect of challenge stressors on job performance via meaningfulness.

Hypothesis 6. Competence moderates the curvilinear indirect effect of challenge stressors on job performance via exhaustion.

3. METHODS AND RESULTS

We collected data from eight different high technology enterprises in China with 407 currently enlisted employees representing 67 teams. On average, the mean age of participants was 30.8 (SD= 6.45) and 65.4% of them were male. Participants had 6.33 years of work experience (SD= 6.10), and, for education, 83.6% had a graduate degree or above. The participants were asked to complete an anonymous questionnaire. Generally, results of the survey provided support for hypotheses regarding the curvilinear relationships between challenge stressors and job performance.

Figure 2. The Relationship between Challenge Stressors and Meaningfulness

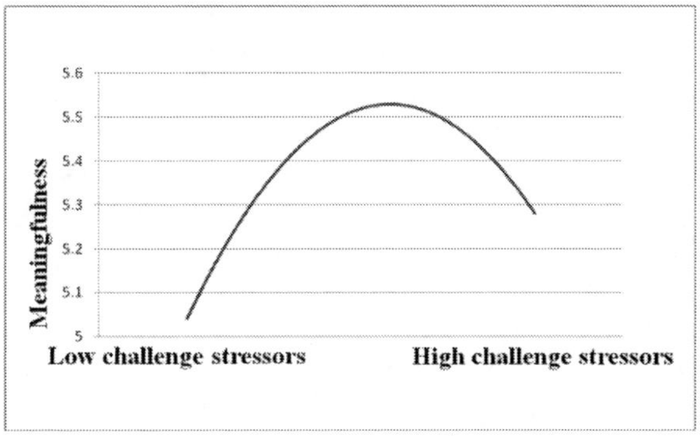

Figure 3. The Relationship between Challenge Stressors and Exhaustion

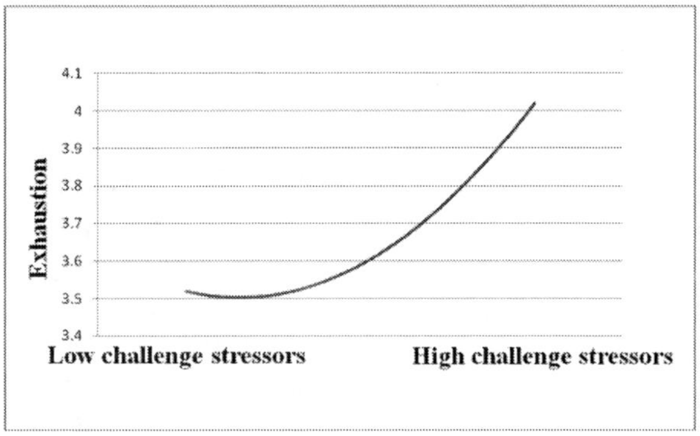

Figure 4. The Relationship between Challenge Stressors and Job Performance

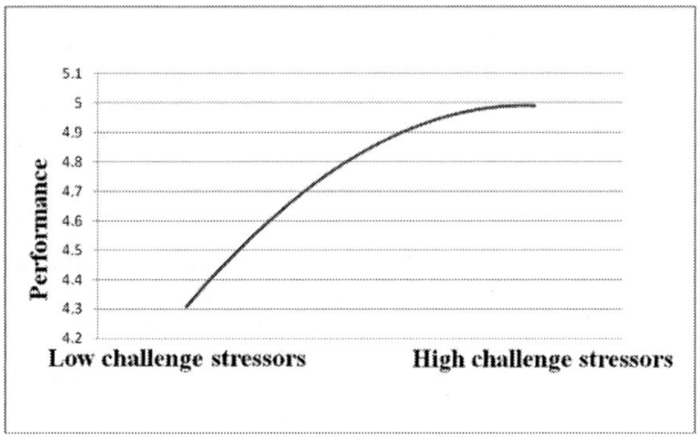

4. CONCLUSIONS AND IMPLICATIONS

The present study developed and tested a moderated mediation to explain how challenge stressors affect job performance. This paper makes several contributions to the existing theory of challenge stressors and job performance. We developed and tested a theoretical model to examine to what extent challenge stressors impact the effects of meaningfulness and exhaustion on individuals' job performance. For theoretical contributions. First, challenge stressors are commonly viewed as beneficial stressors for individuals, which may help to improve job performance. However, previous researches showed mixed results about its effects. Empirical work indicates that challenge stressors have both helpful and harmful effects on the job performance, yet there has been very little conceptual work on it. The study focus on exploring the inverted-U relationship between challenge stressors and job performance, and proposes that excessive challenge stressors may affect performance and act as a stumbling block on way to goal attainment. Second, the study articulates a model that includes meaningfulness and exhaustion to examine how challenge stressors affect the job performance, and the likely performances follow in these two paths. When challenge stressors are within an appropriate range, meaningfulness motivates individual to work hard and make a better performance. However, when challenge stressors keep increasing and exceed an appropriate level, employees tend to achieve few outcomes because of exhaustion. Third, competence helps to direct individuals' appraisal based on the transactional model of stress (Lazarus 1991; Lazarus & Folkman 1984). Using the transactional theory of stress, the study extended the challenge-hindrance stressor framework by proposing that individual competence moderates the curvilinear indirect effect of challenge stressors on job performance.

For practical implications. The study benefits stress management practice and competence development in several ways. First, the findings enrich the content of stress management. This study suggested a new perspective for the research on challenge stressors and its consequences: the inverted U-shape relationship. In this view, challenge stressor is viewed as a contextual variable that may act as a contextual stressor. The research suggests that events appraisals may depend on individual's competence, and emphasizes the importance of positive appraisal for improving performance. Second, the study motivates the engagement of individuals in stress management practices. By amplifying the positive side and mitigating the negative side of challenge stressors, individuals could make a more positive appraisal which benefits the performance improvement and personal growth. Third, our study can also benefit stress appraisal development and training. It is possible to incorporate stressors appraisal ability and competence into the criteria for internal promotions or team design.

REFERENCES

- Barling, J., Kelloway, E.K., and Frone, M. (Eds.). (2005). Handbook of Work Stress. CA:Sage Publications.
- Barrick, M.R., Mount, M.K., and Li, N. (2010). The theory of purposeful work behavior: The role of personality, job characteristics, and experienced meaningfulness.
- Boswell, W.R., Olson-Buchanan, J.B., and LePine, M.A. (2004). The relationship between work- related stress and work outcomes: The role of felt-challenge and psychological strain. *Journal of Vocational Behavior, 64,* 165-181.
- Cavanaugh, M. A., Boswell, W. R., Roehling, M. V., and Boudreau, J. W. (2000). An empirical examination of self-reported work stress among US managers. *Journal of Applied Psychology, 85,* 65-74
- Hunton, J. E. and Price, K. H. (1997). Effects of the user participation process and task meaningfulness on key information system outcomes. *Management Science, 43(6),* 797-812. http://search.proquest.com/docview/213235564?accountid=12660.
- Jackson, S.E. and Schuler. R.W.S. (1985). A meta-analysis and conceptual critique of research on role ambiguity and role conflict in work settings. *Organizational Behavior and Human Decision Processes, 36,* 16-78.
- Jon L. P., Stephen, A. R., and Susan M. (1991). Employee ownership: A conceptual model of process and effects. ACAD MANAGE REV January 1, 1991 16,1 121-144; doi:10.5465/AMR.1991.4279000
- Lazarus, R. S. (1991). Progress on a cognitive–motivational–relational theory of emotion. *American Psychologist, 46,* 819–834.
- Lazarus, R. S. and Folkman, S. (1984). Stress, appraisal, and coping. New York: Springer.
- Lee, R. T. and Ashforth, B. E. (1996).Ameta-analytic examination of the correlates of the three dimensions of job burnout. *Journal of Applied Psychology, 81,* 123–133.
- Leiter, M.P. and Maslach, C. (1988). The impact of interpersonal environment on burnout and organizational commitment. *Journal of Occupa- tional Behavior, 9,* 297–308.

- LePine, J. A., LePine, M. A., and Jackson, C. L. (2004). Challenge and hindrance stress: Relationship with exhaustion, motivation to learn, and learning performance. *Journal of Applied Psychology, 89,* 883-891.
- Lepine, J. A., Podsakoff, N. P., and Lepine, M. A. (2005). A meta-analytic test of the challenge stressor-hindrance stressor framework: An explanation for inconsistent relationship among stressors and performance. *Academy of Management Journal, 48(5),* 764–775. http://doi.org/10.5465/AMJ.2005.18803921
- Lin, W., Ma, J., Wang, L., and Wang, M. (2015). A double-edged sword: The moderating role of conscientiousness in the relationships between work stressors, psychological strain, and job performance. *Journal of Organizational Behavior, 36,* 94-111. http://doi.org/10.1002/job.1949
- Liu, C., Liu, Y., Mills, M.J., and Fan, J. (2013). Job stressors, job performance, job dedication, and the moderating effect of conscientiousness: A mixed-method approach. *International Journal of Stress Management, 20(4),* 336–363. http://doi.org/10.1037/a0034841
- Maslach, C., Schaufeli, W. B., and Leiter, M. P. 2001. Job burn- out. *Annual Review of Psychology, 52,* 397-422.
- Pearsall, M. J., Ellis, A.P.J., and Stein, J.H. (2009). Coping with challenge and hindrance stressors in teams: Behavioral, cognitive, and affective outcomes. *Organizational Behavior and Human Decision Processes, 109,*18−28.
- Podsakoff, N. P., LePine, J. A., and LePine, M. A. (2007). Differential challenge stressor-hindrance stressor relationships with job attitudes, turnover intentions, turnover, and withdrawal behavior: A meta-analyses. *The Journal of Applied Psychology, 92,* 438−454.
- Richard, M. S. and Richard, T. M. (1977). The motivational properties of tasks. ACAD MANAGE REV October 1, 1977 2, 4 645-658;doi:10.5465/AMR.1977.4406744
- Wallace, J. C., Edwards, B. D., Arnold, T., Frazier, M. L., and Finch, D. M. (2009). Work stressors, role-based performance, and the moderating influence of organizational support. *Journal of Applied Psychology, 94,* 254–262. doi:10.1037/a0013090
- Webster, J.R., Beehr, T. A., and Christiansen, N. D. (2010). Toward a better understanding of the effects of hindrance and challenge stressors on work behavior. *Journal of Vocational Behavior, 76(1),* 68–77. http://doi.org/10.1016/j.jvb.2009.06.012
- Webster, J.R., Beehr, T.A., and Love, K. (2011). Extending the challenge-hindrance model of occupational stress: The role of appraisal. *Journal of Vocational Behavior, 79,* 505-516.
- Kahn, W.A. (1990). Psychological Conditions of Personal Engagement and Disengagement at Work. ACAD MANAGE J December 1, 1990 33(4), 692-724; doi:10.2307/256287
- Wright, T.A. Kahn, and Cropanzano, R. (1998). Emotional exhaustion as a predictor of job performance and voluntary turnover. *Journal of Applied Psychology, 83,* 486-493.

Ecology, Environment, and Energy

Field Observation and Wind Tunnel Experiment of Sea Spray Size Distribution under Strong Wind Condition

Hiroki Okachi
Bachelor, Student of Hokkaido University, Japan

1. INTRODUCTION

Associated with climate change, there are increasing concerns about global warming, since, aside from rising sea levels, it affects the frequency and intensity of storms and other weather events.

In 2014, Typhoon No. 26 was formed near the Mariana Islands and grew in intensity on its way over the sea before hitting Japan. It resulted in a total of 173 casualties and missing people and numerous properties destroyed. In the future, it is expected that in the future, the number of tropical cyclones per year will decrease. However, their intensity will increase 1.

There are several studies about the effects of sea spray on the development of typhoons. Once spray droplets which have substantial masses are formed, they are accelerated and extract momentum from the wind. In this process, spray droplets take part on the important process of exchanging momentum at the air-sea surface 2. Typhoon models usually do not include the effect of spray droplets because little is known about spray generation and its effect on the momentum exchange around the air-sea surface. Due to the intensification of typhoons, the generation of spray droplets and its multiple effects on the air-sea surface need to be clarified in order to improve the prediction of the overall development process of typhoons. Correct observational data of spray droplets are therefore needed.

Veron el al. (2012) performed experiments in a wind-wave tank where winds with three different high speeds were generated. In this study, the observation revealed the presence of large drops with diameter of more than 1.5 mm that is not accounted in theoretical estimates of the spray generation. Furthermore, a previously unknown generation mechanism was also observed: liquid sheets form at the crests of breaking waves and are subsequently inflated by the airflow. Eventually the sheets breakup and generate small droplets.

This study reports a marine observation at an offshore tower located in The Pacific Ocean near the coast of Wakayama Prefecture in Japan, where disdrometers (Parsivel) were used. In addition, we report an experiment on an open-channel with a wind tunnel where a disdrometer (SPC) was used. The purpose of this study is to reveal the sea spray size distribution under strong wind condition through direct measurements.

The marine observation and the experiment are described in Section 2. In Section 3 their results are reported. Section 4 provides the conclusions.

2. MEASUREMENT OF SEA SPRAY

2.1 Marine Observation

The disdrometer can detect the diameter and velocity of a droplet within the ranges 0.25mm to 25mm and 0.2m/s to 20m/s, respectively. From these data, diameter spectrum distribution, type of rain, rain intensity and reflection intensity are estimated. Two disdrometers were installed, one at 10m and the other at 15m above sea level. As typhoon wind speed increases, it is expected that the disdrometers can detect more spray droplets. The observation periods were from 21st August 2013 to 15th November 2013, from 21st August to 31st October 2014 and from 21st August 2015 to 31st October 2015. In the first period, typhoons No.18, 24 and 25 passed thorough near the observation tower. In the second period, No.18 and 19 passed.

Meteorological observation equipment of the Japanese Meteorological Agency (Nankishirahama AMeDAS) located 6 km away from the tower was used to determine the no-rain time. However because AMeDAS reads 0mm/h when rain is under 0.5mm/h, c-band radar with a mesh grid size of 1km of JMA was also used to determine the no-rain time. In addition, as the disdrometers were installed on the south side, droplets from the tower can be detected when north wind blows. We then extracted South-wind time by an anemoscope installed at the tower.

2.2 Experiment

Experiments were performed in an open-channel with a wind tunnel. The channel features a piston-type wave generator. The channel is 27 m long, 1.5 m high and 0.6 m wide (see Figure 1). The water level was maintained at 90 cm. The disdrometer (SPC) can detect droplets within the 1.00 mm. The disdrometer was installed at a distance of 11.3 m from the blast gate. The still-water depthis set as 90 cm. The height H from the still water is H=60 cm at the top of the tunnel. The experiment was performed with the SPC installed at varying heights H=18, 20, 22, 24, and 26cm. The rotation number of the blast machine Ω, was set at Ω = 575, 560, 545, 530, 515, 500, 485, and 470rpm. Each measurement time was 10, 20 and 40 minutes.

Figure 1. Schematic Diagram of the Open-Channel with Wind Cover

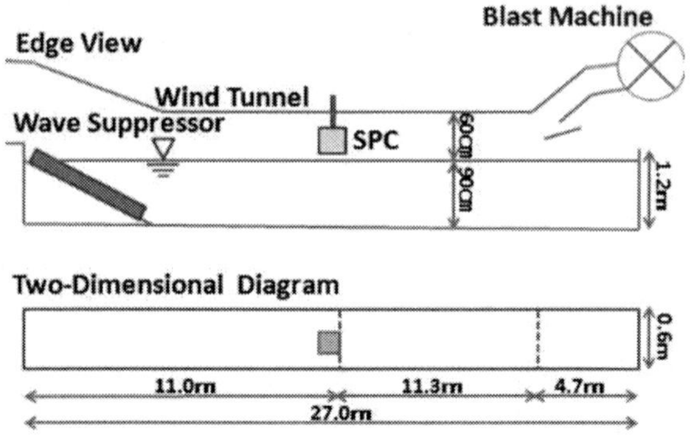

3. RESULTS

Figure 2 shows the sea spray drop size distribution at 15m height from the data of no-rain time in the marine observation.

Figure 2. Number of Particles for each Diameter at 15m Height from the NO-rain Time of Marine Observation

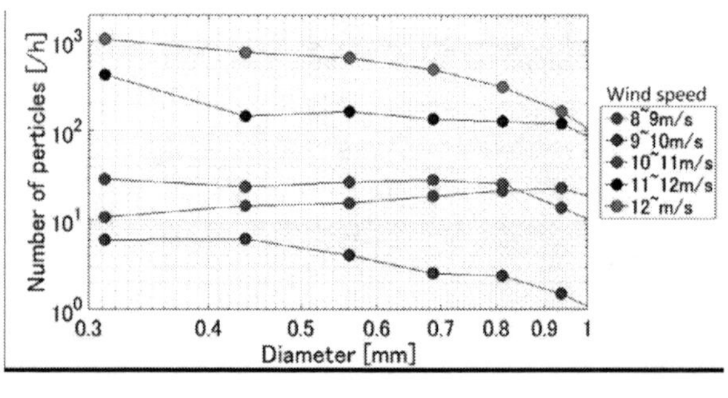

Figure 3 shows that the inclination is similar to the experimental results of F. Veron et al. (2012). As wind speed increases to 24.1m/s, the number of droplets for large diameter increases. The mass concentration becomes constant as the wind speed increase more than 24.1m/s.

Figure 3. Spray Drop Concentration Function Compared with the Experimental Results of F. Veron et al. (2012)

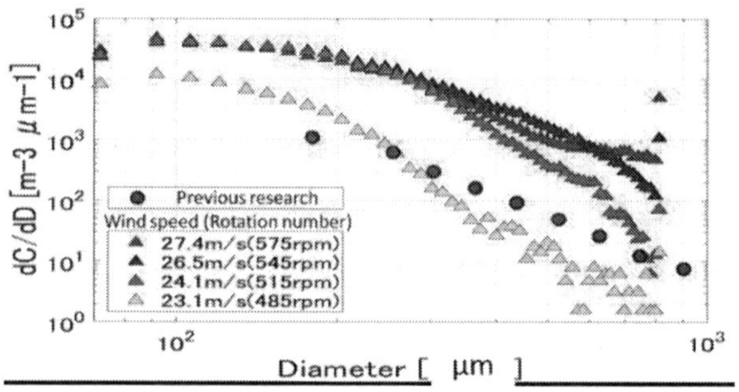

Figure 4. Mass Concentration vs. Wind-Wave Reynolds Number

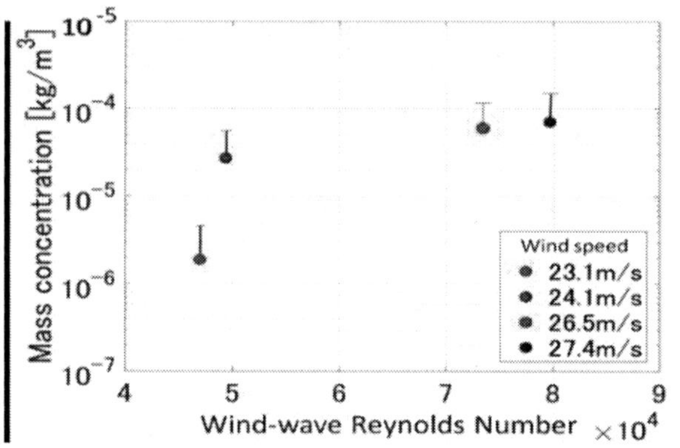

Windsea Reynolds number is an appropriate non-dimensional parameter for characterizing the observational data from laboratories and field3. The windsea Reynolds number is defined as:

R_B=u_*/(vσ_ρ)

where v is the air kinematic viscosity, σ_ρ is the angular frequency at the wind sea spectral peak, and u_* is the air friction velocity. As windsea Reynolds number increases to, the mass concentration increases rapidly.

4. CONCLUSION

We have performed a marine observation and an experiment. Both results indicate that the disdrometers are appropriate to detect sea spray droplets.

As for the experiment, the number of droplets increases sharply when the wind speed increases to 24.1m/s. As the wind speed increases to 24.1m/s or above, the rate of droplets with large diameter increases.

REFERENCES

- HP of Ministry of Land, Infrastructure, Transport and Tourism of Japan.
- Andreas, E. L. (2004). Spray stress revisited. Journal of physical oceanography, 34(6), 1429-1440.
- Toba, Y. et al. (2006). Similarity and dissimilarity in air-sea momentum and CO2 transfers: the nondimensional transfer coefficients in light of windsea Reynolds number. Atmosphere-Ocean Interactions 2, 53-82.
- Veron,F. et al. (2012). 'Sea spray spume droplet production in high wind speeds.' Geophysical Research Letters, 39/16.

Influence of Snow Initialization on the Sub-Seasonal Hydro-Meteorological Fields

Atsushi Nunokawa
Master Course Student, Graduate School of Engineering, Hokkaido University, Japan

1. INTRODUCTION

Previous studies showed that snow information, such as snow amount or snow melting process in global scale, have important roles in hydrological cycle. Koster et al. (2010) suggested that the skill in streamflow forecasts in the Mississippi river in spring time can increase by using snow amount and soil moisture initializations, and especially snow amount contributes to the increase of skill. Arai and Kimoto (2005) showed that when the surface temperature over Siberia is higher in April, atmospheric blocking events over northeastern Siberia and the Sea of Okhotsk in May and June, occur more frequently than climatology. They suggested that spring time long memory of the surface condition, for example, snow cover which begins to melt in April, over Eurasia continent influence the upper atmospheric circulation and subsequent early summer blocking activities.
Remote sensing enables us to discuss geophysical distribution of snow cover in global scale. However it is difficult to estimate snow depth and its density. This study investigates the influence of snow initialization on the sub-seasonal hydro-meteorological fields by the land surface variables reproduced by a global Land Surface Model (LSM) with observation data. The contribution of snow initialization for hydro-meteorological forecast skill is also examined with an Atmospheric Circulation Model (AGCM).

2. METHODOLOGY

2.1 Land Surface Model

An LSM used in this study is Minimal Advanced Treatment of Surface Interaction runoff (MATSIRO; Takata et al. 2003). The MATSIRO is the land surface scheme of an AGCM which expresses all the important processes for the water rand energy exchange between land and atmosphere. This model reproduces these 6 processes below: 1) transpiration control by open and close of stomas; 2) radiation process in vegetation group and heat and water transfer by atmospheric vortex motion; 3) cover and evaporation of precipitation by vegetation; 4) snow on the ground and vegetation, heat conduction in snow and refreeze of melt water, aging effect of snow; 5) out flow of water on the ground surface and underground considering gradient of land; 6) heat and water transfer, change of phase of soil moisture (soil frozen process).

Atmospheric forcing variables which were given to MATSIRO to estimate land surface variables, e.g. snow amount, snow albedo or soil moisture, are based on the atmospheric reanalysis data provided by Japanese Meteorological Agency (JMA) Climate Data Assimilation System (JCDAS). Clouds cover, radiation, surface pressure, precipitation, specific humidity, wind velocity and temperature were used to reproduce the land surface condition. Obtained land surface variables were given to an AGCM for forecast simulations to investigate potential predictability.

2.2 Atmospheric General Circulation Model

The AGCM used in this study is the Model for Interdisciplinary Research on Climate (MIROC), which is the coupled general circulation model consist of AGCM ver5.7, developed by the Center for Climate System Research at Tokyo University (CCSR) and the National Institute for Environmental Studies of Japan (NIES), and MATSIRO. It consists of primitive equation system and acquires prognostic variables each time steps by solving governing equations i.e. equation of motion, the equation of continuity, thermodynamic relation and the equation of hydrostatic pressure.

2.3Atmospheric Data

National Centers for Environmental Prediction (NCEP), National Centers for Atmospheric Research (NCAR) reanalysis data was used for atmospheric initialization for the forecast simulations. This data set has 4-times daily time step, $2°\times2°$ resolution and 28 sigma levels. 10-members of ensemble with 3-hours perturbation were prepared for the ensemble forecast simulations.

2.4 Forecast Simulation

To investigate the effect of snow initialization to the sub-seasonal hydro-meteorological forecast, 2 types of experiments were performed in every March (30 days forecast simulations from March 1st to March 30th) between 1998 and 2007.

Table 1. Experiment Design

Exp.	Atmos. Init.	Land Init.	Num. of Exps.
Experiment 1	Reanalysis (3-hr perturbations)	Realistic	100 (10 years * 10 ens.)
Experiment 2	Reanalysis (3-hr perturbations)	Randomly Chosen	100 (10 years * 10 ens.)

In Experiment 1, atmospheric initialization is the NCEP/NCAR reanalysis data with 10 ensemble members. The land surface variables, correspond to the target year was applied to the 10 ensemble members as the land initialization. Note that the snow initialization was applied a scaling technique to adjust the climatological characteristics of the AGCM. In Experiment 2, the equivalent 10 sets of atmospheric initializations to Experiment 1 was given. However the 10 sets of land surface initializations randomly chose the land surface variables from different 10 years between 1998 and 2007 on March 1st.

Experiment	Atmos. Init.	Land Init.	Num. of Exps.
Experiment 1	Observed (10 members)	Real	100
Experiment 2	Observed	Random	100

3. RESULTS

I compared observed data and the result of 2 types of experiments. The difference between observation data anomaly and simulation anomaly becomes smaller in Experiment 1 in both 2 meter height air temperature and sea level pressure. In the result of correlation coefficient in each grid, the coefficient between observed data and Experiment 1 is higher than that of observed data and Experiment 2. The snow initialization can reduce the gap between the forecast result and the observation not only over land but ocean.

4. CONCLUSIONS

This study examined total 200 patterns of experiments (10 years × 10 ensemble simulations × 2 experiments) to investigate the influence of snow initialization on the sub-seasonal hydro-meteorological fields by using atmospheric observed data and land data reproduced by the LSM. The result showed that the spread among the ensemble members in Experiment 1 is smaller than that of Experimental 2. This shows that snow initialization has a potential to contribute to the increase of forecast skill in this 1 month simulation, which beyond the time scale of atmospheric memory. In addition, this study suggests that the snow initialization increase the relativity between observed data and the experiment result, not only on the land, but ocean.

Figure 1

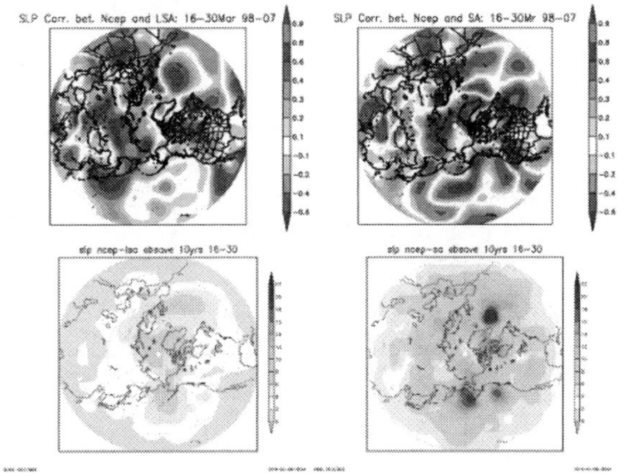

Upper left: Correlation coefficient between anomaly of observed data and anomaly of Experiment 1 result, in day 16-30 (16Mar-30Mar) in each grid in sea level pressure in 1998.
Upper right: Correlation coefficient between anomaly of observed data and anomaly of Experiment 2 result, in day 16-30 on each grid in sea level pressure in 1998.
Lower left: Absolute average error between observed data and the result of Experiment 1 in day 16-30.
Lower right: Absolute average error between observed data and the result of Experiment 2 in day 16-30.

REFERENCES

- Koster, R.D., Mahanama, S.P., Livneh, B., Lettenmaier, D.P., and Reichle, R.H. (2010). Skill in streamflow forecasts derived from large-scale estimates of soil moisture and snow. *Nature Geoscience, 3(9),* 613-616.
- Arai, M. and Kimoto, M. (2008). Simulated interannual variation in summertime atmospheric circulation associated with the East Asian monsoon. *Climate dynamics, 31(4),* 435-447.
- Takata, K., Emori, S., and Watanabe, T. (2003). Development of the minimal advanced treatments of surface interaction and runoff. *Global and planetary Change, 38(1),* 209-222.
- Kalnay, E. et al. (1996). The NCEP/NCAR 40-year reanalysis project. *Bulletin of the American meteorological Society, 77(3),* 437-471.
- Barnston, A.G., Chelliah, M., and Goldenberg, S. B. (1997). Documentation of a highly ENSO-related SST region in the equatorial Pacific. *Atmosphere-ocean, 35(3),* 367-383.

Design, Technology, and Innovations

The Road to the Technology Singularity

Phillip Mathew
Bachelor of Economics, Macquarie University, Australia

INTRODUCTION

With time being infinite and technology growth consistently exponential; *to what end is technology going to grow?* Logically, the answer would be an event characterized by either time or technology growth coming to a halt. However as neither of these scenarios are recognized as potential reality, the popularly accepted answer is a conceptual event; the Technology Singularity.

The purpose of this paper is to advocate championing the appropriate development of technology so that the speed at which we are getting to the technology singularity can be accelerated. In order to avoid blurring this purpose, the body will be divided into three distinct parts. Firstly, it will provide a conceptual understanding of what the Technology Singularity is. Next, it will unpack technology through the lens of computing capacity, interconnectivity and ideas, illustrating the positive impact it has brought. Finally, I will conclude with a notion that will provoke thoughts of change in attitude and perception of what may lie in the near future.

1. WHAT IS THE TECHNOLOGY SINGULARITY?

As the opening question may have implied, the Technology Singularity is not an end to technological growth, but rather a critical moment in time, determining to what extent technology will be intertwined with our future. Kurzweil (2013)[1], an academic author in the study of technology and the future, effectively explains the Technology Singularity through the collective implications of two key ideas:

1.1 Technology Grows Exponentially

In 1965, Gordon Moore, co-founder of Intel Corporation, released an article suggesting that the number of components able to fit on an integrated circuit board would begin to consistently to double every decade implying computing capacity, a pillar of technology, would begin growing at rapid speeds[2].

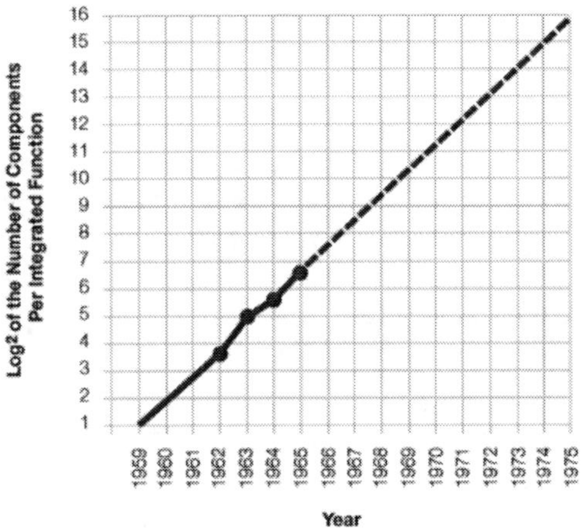

Despite changing his perspective of this exponential rate to an even faster two year cycle, the underlying theory of exponential technological growth became popularly labeled 'Moore's Law' and over time has been commonly considered the roadmap of technology growth.

Despite some suggestions this law may saturate in the next decade, to assure a conceptual understanding of the technology singularity, Moore's law is to be held true throughout this paper.

1.2 Human Intelligence Will Be Surpassed by Machine Intelligence

As the rate of technology growth continues to accelerate, Kurzweil puts forward that eventually machine intelligence will eclipse human intelligence.[3] However in order for this to happen, there are two milestones that must be passed. The first step is the ability to replicate the physical functions of the brain which essentially has already been completed with medical practices already including the use of neural implants; we have the technology to reverse engineer and create parts for the brain. The second step is the replication of human intelligence. This is more referring to facets of the mind such as emotion, conscience or logic. Through advancing studies of areas such as genetic sequencing, evolutionary algorithms and programmatically designing artificial intelligence, this step is on the verge of its completion.

[1] *R Kurzweil, Center for Strategic & international Studies: Ray Kurzweil on the Singularity, 2013.*
[2] *G G Moore Cramming more components onto integrated circuits. Research and Development Laboratories, Fairchild Semiconductor division of Fairchild Camera and Instrument Corp, 1965.*

[3] *R Kurzweil, The Coming Merging of Mind and Machine. Scientific American Special Report on Robots, 2008.*

Once both steps have been taken, *'Intelligent creations will soon eclipse us – and then their creations will eventually eclipse them'.*[4] The implications of such an achievement, amplified billions of times, will make the potential reality difficult to imagine; a characteristic of the singularity.

2. THE IMPACT OF TECHNOLOGY

As primitive a discovery that fire begat light and heat, right up to the present day where artificial intelligent machines begets empathetic decisions, since the beginning of time the huge growth in technology has had largely positive impacts. Today, we have evolved into the digital age. This is an era where information and data, and the digital platforms through which they flow, sit at the core of our global blueprint.

To take a deeper dive into technology and its impact, we can view it in three lenses; Computing Capacity, Interconnectivity and Ideas.

2.1 Computing Capacity

Growth in computing capacity has allowed for big leaps in many industries. Examples of this are science and mathematics and new found abilities to reach new heights of knowledge, with atomically intricate hypotheses and previously complex calculations now being comprehensible.

In 2013, Adam Spencer, an Australian media profile and qualified academic in the field of mathematics, delivered an insight into mega-prime numbers[5]. As part of his presentation, he pointed out that prime numbers were used as tests for limits on computing capacity. He went to on to show that prior to the start of the computer age (circa. 1950s), the largest known prime was 44 digits long. However through significant improvements in compute ability, the largest known prime today is 17,425,170 digits long. On the graph shown[6], each blue dot represents the number of digits in the largest prime number known in each respective year. Prime numbers were reflective of computing capacity as they were used as tests for limits on computing capacity[7], so this graph makes

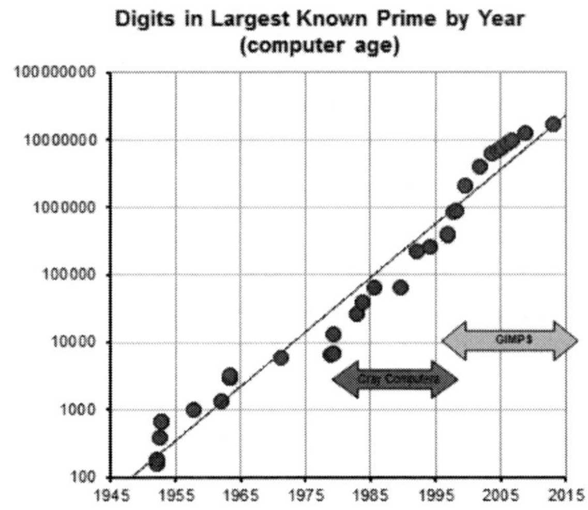

evident that as computing capacity was growing, so was mathematics.

Evolutionary Computation is another example of how growth in computing capacity can and will continue to provide significant opportunity. Evolutionary Computation is the process of 'abstracting evolutionary principles (Darwinism and natural selection) into algorithms that are then used to search for optimal solutions to a problem'.[8]

The three dimensional graph[9] shown is a simple example of how this computation occurs. The sample population makes up all possible combinations of trait 1 and trait 2. Every candidate in the population is run through all permutations of algorithms, representing all possible external influences, with the resulting

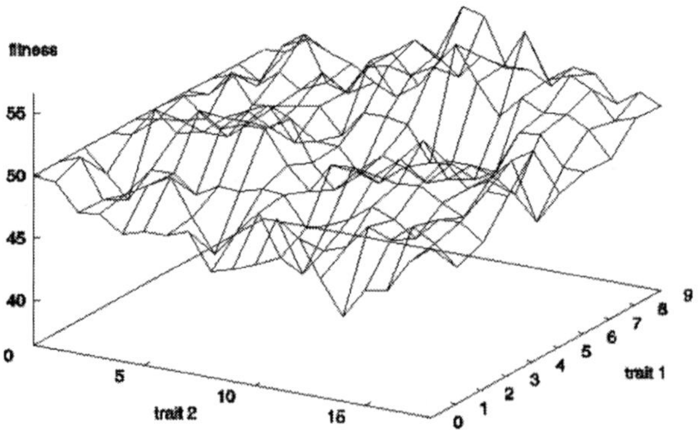

[4] *R Kurzweil, The Coming Merging of Mind and Machine. Scientific American Special Report on Robots, 2008.*

[5] *A Spencer, Adam Spencer: Why I feel in love with monster prime numbers [Video file], Retrieved from www.ted.com/talks/adam_spencer_why_i_fell_in_love_with_monster_prime_numbers, 2013, February.*

[6] *Ch Caldwell, Department of Mathematics and Statistics, University of Tennessee: Digits in Largest Known Prime by Year. https://primes.utm.edu/notes/by_year.html, 2012.*

[7] *A Spencer, Adam Spencer: Why I feel in love with monster prime numbers [Video file], Retrieved from www.ted.com/talks/adam_spencer_why_i_fell_in_love_with_monster_prime_numbers, 2013, February.*

[8] *H S Talib, An Introduction to Evolutionary Computation, Department of Computing and Information Science, Queens University.*

[9] *E A Eiben and E J Smith, Introduction to Evolutionary Computing, Slides for an EC Course [ppt file], Retrieved from http://www.cs.vu.nl/~gusz/ecbook/ecbook.html.*

fitness level (optimism and quality of solutions) identifying the fittest solution to the problem.

Placing this into the aforementioned context of machine intelligence surpassing human intelligence, this process of computation would allow for artificial intelligence to follow the evolution of human intelligence, improving on any inefficiencies and essentially giving artificial intelligence the platform to surpass human intelligence and essentially becoming as real.

Realistically, such an outcome would require mega-computational capacity justifying investment of time and resources to grow computing capacity and allow for this type opportunity to come to fruition. The impact would be felt in globally relevant fields including medicine, sciences & technology.

2.2 Interconnectivity

Interconnectivity has been at the core of the impact that modern technology has made. Most notably in today's digital era, our global infrastructure is built on the back of being globally connected. Economic and Political Relations are all handled without leaving a room, International Commerce and Business transact billions of dollars across the globe, Social Networks connect people on opposite sides of the world and Academic Research being a collaboration of specialists from different parts of the world; All these globally meaningful realities are made possible because of the interconnectedness of everything.

Testament to the impact interconnectivity has, was the sequence of statements that Indian Prime Minister Narendra Modi delivered at his interview, hosted by CEO Mark Zuckerberg at the Facebook headquarters earlier in the year. Modi emphasized that *'The strength of social media today is that it can tell governments where they are going wrong. It can stop them moving in the wrong direction'.* He followed this by speaking about his personal use of social media and how on wishing the Chinese Prime Minister for his birthday that it became national headlines and how wishing the Israeli Prime Minister in Hebrew for an occasion prompted him to respond in Hindi. To this he said *'who would have thought that this could be a form of diplomacy...this (social media) is a new face of diplomacy'.*[10]

[10] *N Modi, Townhall Q&A with PM Modi and Mark Zuckerberg at Facebook HQ in San Jose, California, retrieved from: https://www.youtube.com/watch?v=1-OFfyf7aoA, 2015.*

2.3 Ideas

Despite all the progress visible in the two previous scopes, it is the innovative ideas of putting it all together that see the greatest achievements. One of the most powerful examples of this is the Microsoft Corporation's product, Skype Translator. It is effectively a video calling platform, leveraging an extensively trained translating program, to allow people from around the world to communicate even if no common language is available. Person A will speak in one language into the microphone, the translator will analyse what was said with all its accents and nuances, and then relay the message in Person B's native language – in real time. The implications of such an idea go beyond just connecting people. It amplifies out into the realms of education and business. It means that educational institutions can engage students and teachers from different parts of the world to learn and experience things they otherwise would not have learnt or experienced. It means that international business dialogue will be free flowing with language no longer an obstacle. Ideas of such magnitude have been at will continue to be at the core of the impact technology will have.

3. A NOTION FOR THOUGHT

As we continue along the infinitely long timeline, technology will continue to grow. As this growth comes to fruition we can either embrace it or push it away. Inevitably, we are going to reach the Technology Singularity but it is in our control as to how long it will take.

We need to forget the failed attempts by fiction to displace confidence in the concept of man merging with machine and push aside our humanly flaw to desire being better than one another. We must work in the best interest of the global community. We must embrace and support the inevitable growth of technology and welcome machine to learn at the side of mankind so that one day it will surpass human intelligence, overcoming the limiting factors of human life, and only once this happens will we all bear witness to the start of a new beginning.

REFERENCES

- Kurzweil, R. (2013). Center for Strategic & International Studies: Ray Kurzweil on the Singularity.
- Moore, E.G. (1965). Cramming more components onto integrated circuits. Research and Development Laboratories, Fairchild Semiconductor division of Fairchild Camera and Instrument Corp.

• Kurzweil, R. (2008). The coming Merging of Mind and Machine. Scientific American Special Report on Robots

• Caldwell, Ch. (2012). Department of Mathematics and Statistics, university of Tennessee, Digits in largest known prime by year, retrieved from https://primes.utm.edu/notes/by_year.html.

• Talib, H. S. Department of Computing and Information Science, Queens University. An Introduction to Evolutionary Computation.

• Eiben, A.E and Smith, J.E. Introduction to Evolutionary Computing Introduction. Slides for an EC course retrieved from http://www.cs.vu.nl/~gusz/ecbook/ecbook.html.

• Spencer, A. (2013, February). Adam Spencer: Why I feel in love with monster prime numbers [Video file]. Retrieved from http://www.ted.com/talks/adam_spencer_why_i_fell_in_love_with_monster_prime_numbers.

• Modi, N. (2015). Townhall Q&A with PM Modi and Mark Zuckerberg at Facebook HQ in San Jose, California. Retrieved from: https://www.youtube.com/watch?v=1-OFfyf7aoA.

All Technology is Obsolete When It Is Not Sustainable: The Two Sides of Globalization

Jesús Alexis García Piña
Administration Student, Universidad Autónoma Metropolitana Campus Iztapalapa, México

INTRODUCTION

If we talk about innovation and technology the first stuff that arrives at our minds it´s small devices that will help us to make everything easy and to spare time.

The corporations that control the international markets provide the world with an enormous amount of devices and services that help the costumers to satisfy a necessity but are these necessities essential to the human welfare?

We are developing technologies and some other innovations to sell things and this will lead us to a great profit. The corporations also can have a profit if they develop goods but with the additional value that requires the planet, we are talking about sustainable products and some strategies that will help to develop this kind of goods.

This is a big step in technology if we develop the majority of our products and services thinking in how really help the world and the future of the humanity we can satisfy the necessity of the people and looking into the future generations. We'll be able to create not just the cutest devices and the idea to re-use them after their useful life was ended. In that manner the costumer will accept that it´s very important to take actions to face ecological problems generated by humans. These changes can be reach by the application of the technology and new ideas that lead the old products adding innovations on them, looking in their value chain.

Innovations also can be applied on strategies that can work with the right technology this kind of solutions can be reach looking in small details into the production or in other parts when we are developing. For that reason it's a duty to think in these actions when we´ve the technology and of course the money to invest not only in goods that generate a high profit if companies doesn´t take this kind of awareness they will be responsible of saturate the world with more non-recyclable waste.

BODY

These useful things were created for the human welfare with high-tech and most of the times without workers, doesn´t matter the investment that requires not only money and the quantity of resources to develop that technology. The new corporations use their potential talking about ideas and then they apply any quantity of resources and money to develop any service or products that will help them to raise their profits.

The next step in business it's to make that product cheaper than their competitors. In the moment the corporations search to find new ways to spare money in all their supplies or even at the manufacturing phase.

These two corporative weapons are easy to reach with the help of the globalization that make possible to buy everything with a simple click using the internet. Nowadays we can find everything using this tool but the internet is able to deliver any product from one continent to another, but we never looked the small letters that will be written with red ink.

As fast as we receive our orders and with a lower price we will be happy to make our products we should know how they produce the products and all the materials used to the final product. The same case apply to the costumers that bought the products, all of them wants a cheaper product that will satisfy a necessity and this one should be with a low price and easy to get.

On this process we find a mix of technology and innovation that's necessary to develop things nowadays, as human beings the necessity to create new stuff and show it to any one that works in the same field creating an international competence to have the best technology that will help us to develop faster than the competence and with a lower price.
All the raw materials and workers that helped in some cases in the production or in recollection of materials most of the times doesn´t have the technology this is a kind of joke because we built amazing things with simple ideas but with a simple donation of technology

corporations should have a duty with the world and not only with their customers.

If we have the technology and the power to bring raw materials from anywhere with this global connection a first approximation of the Clegg definition (2007, p. 135) quotes Therborn (2000) is or are the "trends of social phenomena to a scope, impacts and global connectivity or to an awareness globally intertwined between the social partners". This is thanks to the capitalist model which enables the global connectivity exist. Which not only allows it to continue to grow and is on the rise in the proportion in which it is carried over, and that all these new goods and services that are being created allow to remain profitable meet the needs of the globe sharing different kind of technology?

These profitability and the primary purpose of the application of technology is what has always sought and allowed these gains leading to a direct investment in new things or seeking to add value to what already exists (innovation), globalization permits free expansion of goods, services, materials and technologies used for new purposes.

This allows to have a global character (Natalichio 2011) being globalization economic, technological, social and cultural scale process, which involves increased communication and interdependence among countries of the world uniting their markets, societies and cultures, through a series of social, economic and political transformations that give them global reaches.

Profits are partly a direct consequence of the whole process brought by the technologically advances among countries and this creates a demand, at the end globalization brings any technology around the globe allowing a transfer in technologies that will take care of the health of our planet.

Evoking different characteristics (Kellner 2002) globalization is developing a "product" result of the technological revolution and the global restructuring of capitalism in which economic, technological, political and cultural characteristics are intertwined. Thus this process can already take also as a 'product' which is still in critical by stakeholders who have an indirect role in it, but this product is still changing and adapting as well as breaking old models becomes obsolete because of the big technology and systems are still forming progress.

This social-technological-governmental result is a mixture of all the elements that make up but is shaped by the needs and demands of the people who consume it.

We know the rules were defined by nations or companies offering the product or services that we want handled by a third party. Each of these rules and regulations are different in each country and in each product but the major interest is to share green technologies that will permit produce green products maybe not at 100% of the product but at least the process when the product is garbage.

That is why we can talk about a globalized era (Canclini 2002) which will help us to transform any production with a little interest in the global welfare. It is created with the individual idea and passed on to a higher rank and so every moment.

For Ianni (2000) are transnational which have the power and strength to move countries and cause them to suit their needs, another way to pass this global contribution. The impact of globalization affect (Stiglitz 2007) societies and economies, this can be seen in an explicit way as globalization trends and basically seeks to generalize this, creating a trend and innovation need to be unfolded in a society; this generates the change in the same. This makes people in every society to assimilate differently and this will generate a change in people and their lifestyles, leading to changes in behavior are made in the majority the person and the institution. In that way technology allow behaviors, services, materials, and lots of things that make the features of anything be modified to adapt in a sustainable way.

This will mean that is reflected in organizations that fit this and looking to have a competitive advantage in the search for new market niches so they can attract new customers, since the effect of technological advances permit a discussion of the impact in the world welfare.

Once identified that organizations can transmit this message and for their own profits of course allows them an organizational growth and at the time that organizations use it directly as affectation on their economies.

Of course first world economies had always an advantage since they have the latest technology and resources to create it and of course we´re talking about mass production, allowing them to generate the major quantity of trash without a sense of sustainability.

And this gets into the forces driving nations either first world or third world as the impact of globalization and how it will affect (Stiglitz 2007) socially, politically environmental impact. While globalization and technology is not only an international growth (Stern, R.M 2011) as foreign investment, multinationals, integration of world markets and all this so that a flow of financial capital is encouraged. Thus technology has always been present since as said these advances always encourage that globalization is distributed differently, for example with the creation of computers and how we have been using for the creation of programs and databases and it ended with the creation of communication networks that will help us to transfer sustainable conceptions.

The importance of new technologies and the power of global corporations should work in an ecological way restructuring the mass production system and with a new view of point thinking in the future. This part is fundamental because we believe that innovation is only the products that we see in stores but it will work in other phases. Even technological advances allow the creation of new materials for products adding value not only for human daily necessities such technology in the sole that is required for footwear for an athlete when it´s more important a healthy world. Sustainable development (Clayton, A. 1996) is achieved when the needs of the present is known as the commitment that the usability of future generations to their own needs is not compromised. This makes it look optimally use the resources you have right now and which are implemented with the available technology.

That is the hardest part because although the various ecological problems worldwide will causing precisely for this application of technology and in the process of applying these resources are used and exploited at a level that prevents you never regenerate known.

The carrying capacity has been exceeded by far and this makes different resources continue to be used without limiting it, which makes the resort to go extinct and that although we have the knowledge of this situation is the fact that resources they needed at an alarming rate in the daily lives of people not allowed to be carried out a reduction in the use of technology, goods or services. It should be noted that while products are searched through a process of greener production has not yet come to take action to decrease this alarming need to material things, namely the fact that there are solar panels allows a minimum amount of will save energy in a home and multiply that amount of energy for thousands of homes that have the same savings as system energy savings would reach a considerable scale. But the fact is that to create a solar cell the amount of carbon and fossil oil used to create a simple solar cell is alarming it puts into question the 'energy-saving', since for processing this materials and compress them to reach the final product the process has nothing ecological and complexity of this transformation ends up making the final product actually be as friendly to the environment or that it actually helps as such in the future.

The system of sustainability (Clayton, A. 1996) is focused on subsystems or takes account of biological, ecological, social and economic nature. This causes it to become very complex and that the analysis in the creation of products or services is not so easy, because this system is interconnected not only with the technology but with the same knowledge we have of new technology but where is coming.

This is the new challenge is to encourage, promote and raise awareness of sustainability as it allows companies with an interest and not only significant but thoroughly what they are doing not only in their final products but in their production processes and suppliers providing them inputs they require. In this part we face the ideas with the innovation that will be able to reduce waste and toxic parts in the production. Thus if we look at the whole system so thoroughly that is interconnected with more people and companies (using the correct technology), it makes this need to consider these issues are spread in the same way some need other products and the fact that they should have consideration about what we´ve doing and the environmental and social impact that spreads and promotes all this interrelated system.

Another important part that we must bear in mind that this adjustment being coupled to systems to save more in all areas described is the way we treat waste technologies that allow us this ecological transformation both mentally and physically. Production systems that normally supplying the population become "intensive production systems" that because we no longer speak of a controlled population, but a global overpopulation.

The impact of the lifestyle of the world's population is somewhat alarming not only because of the number of individuals that exist in the world and the necessities of life that adversely daily is not only the waste that is generated but the creation service that carries this therefore adapting or creating strategies (Ottman 2011) which reflect this change in behavior in people and organizations, the latter are the ones to be aware of this situation it is necessary and they are the ones that provide different things and services.

This is where technological advances are taking into account as this is a big fight between some ecological processes are expensive and the fact of trying to take account not always lead to more gains in the products, because there will always be sold the product / service to one that is cheaper but the inputs are lower quality or have an effect on the environment and another negative into the society and their future. Onsite there is increasing green products (they have almost no environmental impact) this is a perfect example when you mix technology and innovation.

This innovation involves not only ecological impact but considers adding social value and is of the most important things it brings, and that just takes into actions in one political aspect. Conversely 'green' products are those that are focused on the concern that everything is natural and take into account environmental impact they may have on production. Green consumers (Ottman, 2011) are all these new consumers who have this minimally conscious by concerns everything that is affecting the planet such as global warming or the amount of global waste, they know the technology and new materials helps the planet and consuming them helps to generate a change in proportion to what they consume.

Thus organizations should start promoting these issues of global importance, since it is obvious that these issues should no longer only be seen but perform take action to promote a new way of thinking and this will lead to a new culture as new generations are those that end with an unimaginable degree of involvement. There are segments Green Interest (Ottman, 2011) from resource-conserving.

As are the technologies and the application of this science which allows to get to know the 'green' processes and to measure the impact of global pollution makes plans to decrease are believed knowledge is generated.Thus this technology application cares or has a new effect on the main factors (Ottman 2011) natural materials, production, packaging, distribution, use and waste after use. And all this can be modified to have a positive impact at every stage of any process in which it is spoken. Here is when innovation takes place and introduces new ideas and processes.

One strategy involves the creation of a unique position which is involved in a set of activities (Orsato 2009). This position is the part that makes an approach to a competitive or just advantage to be focused on it.

Strategies for each name makes it easy to identify what you want to do (technological advance), this will depend on the type of organization and the context in which they are placed.

CONCLUSION

If we apply the technology and the innovation not just with the intention to make the life easier for everybody and with the purpose to sustain the things that we are developing, in that way the resources and the awareness for the surveillance and the care of future generations will be our major goal. The knowledge is present nowadays but we are really taking care for the next generations?

Maybe is not about morality but if we apply these smart decisions and take care of all the wastes that we generate day after day this knowledge and the appliance in the technology will create the necessary actions to survive in our days and in the further days erasing the alarming amount of toxic wastes our processes will be friendly with our environment.

These processes should be supported by any corporation doesn´t matter which kind of product they produce but with the major goal to help the life of the planet using sustainable ways of production.

When we transmit these to our clients they will be able to look the labels and to reduce what they consume, it´s about to create a human change with the collaboration of the companies of course they want high profits but even developing this kind of products they will satisfied their objectives.

That's the importance to use high-tech in the right way and not just developing weapons or cutest devices that only contain toxicity after their useful life, this high-tech should be applied in some other process also for societies that doesn´t have the resources as first world country has.

REFERENCES

- Anthony Clayton, N. R. (1996). Sustainability a system aproach. United States of America: Westview Press.
- Canclini, N. G. (2002). Globalization (1st editioned.). México: Paidós.

- Ianni, O. (2000). Scielo. Obtenido de http://www.scielo.org.bo/scielo.php?script=sci_arttext&pid=S1815-02762003000200004&lang=pt
- Kellner, D. (2002). BIDI UAM. Obtenido de http://www.bidi.uam.mx:8331/V/LLNA7ITKBH92H7FYP2LC9G692RDIGQ74G9H35BX67N6U7RQ591-39615?func=find-db-6&short-format=002&set_number=000434&set_entry=000007&format=999
- Natalichio, R. (2011). SCIELO.From http://www.scielo.org.co/scielo.php?script=sci_arttext&pid=S1909-24742011000100001&lang=pt
- Orsato, R. (2009). Sustainability strategies when does it pay to be green? Great Britain: Palgrave Macmillan.
- Ottman, J. A. (2011). The New Rules of Green Marketing. UK: Greenleaf Publishing.
- Stern, R.M. (2011). Comparative Advantage, Growth, and the Gains from Trade and Globalization. USA Michigan: World Scientific.
- Stewart Clegg, S.T. (2007). Estudios Organizacionales y Asuntos Globales. México D.F.: Comité Editorial del Departamento de Economía.
- Stiglitz, J.E. (2007). Making Globalization Work. U.S.A.: Norton Paperback.

Education and Youth

Session I: Facing the Challenges

Preventing Radicalization in the Digital Era

Nourah Moahmmed Al Jammaz
Student, Al Yamamah University, Saudi Arabia

ABSTRACT

This paper highlights the situation of today's youth regarding the prevention of terrorism. Currently, our youth is being targeted worldwide by terrorism's most evil conspirators. Strategies have been formulated by terrorists to control and recruit young people into committing acts of violence, using social media. It has become apparent that the threat of manipulating new terrorists through the World Wide Web is indeed real and it requires constant monitoring. All must be done to ensure that the younger generation is of sound mind when it comes to recognizing the risk of terrorism. Therefore, educational efforts must be made to instill positive values into today's youth population. To help prevent the spread of terrorist activities, the international governments must fulfill its role of protecting the youth while combating the strategies mentioned above. Education is vital to the awareness of these concerns, and to stopping behavioral problems while reinforcing personal and social development directed toward the highest potential. My research into this ongoing problem was done to better understand the current landscape of terrorism and its youth involvement, as well as to determine solutions on how to best prevent continuous online recruitment. Incorporating a cross-section of scholarly articles and news, this paper will examine how patterns of socialization, technology, history, geopolitics, poverty and war have resulted in the exploitation of vulnerable youth to the benefit of the cause of terrorism.

1. INTRODUCTION

1.1 Context: Terrorism Today

Terrorism is a very difficult problem for the world to confront. The force of hatred fueling acts of destruction across the globe has persisted enduringly, leaving a tragic casualty record with a consistent pulse. The scope of terrorism is decidedly international in its reach, with countries from the first to the third world witnessing acts of extremist violence perpetrated by both locals and travelers alike, independently influenced or affiliated with groups. Degrees of organization are varied, but groups such as Daesh/ISIS/ISIL (whose stated objective is to establish a caliphate spanning multiple countries) and African pseudo-affiliate Boko Haram have made headlines for their aggression and momentum. Their growing numbers and increasing reach has been aided by the ability to achieve modicums of stability in economically weak or politically unruly regions, such as war-torn Syria, or Libya (whose post-Gaddafi power vacuum left room for new groups to take power). The ongoing Israel-Palestine conflict, the American invasions of Iraq and Afghanistan, Iran's contested nuclear program, and Egypt's internal political upheaval comprise surrounding issues relevant to the state of political turmoil in regions affected by terrorist activity.

Figure 1. Indicates the International Presence of Terrorism within this Generation[1]

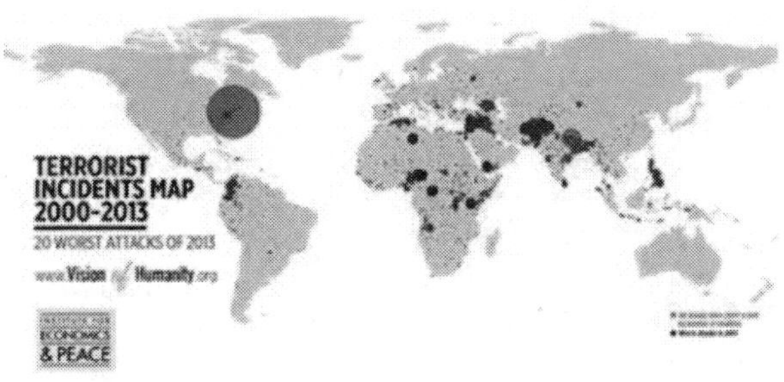

Figure 2. Displays Areas Where Government Overthrow, Protests, Civil War or Militant Conflicts Have Taken place within the Recent Past. These Changes Have Nurtured a Geopolitical Environment that Accommodates the Burgeoning and Continuous Facilitation of Terrorism. [2]

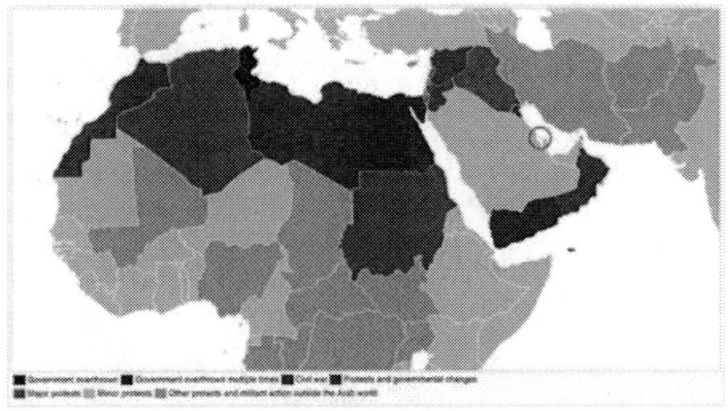

[1] *Z Beauchamp, 33 Maps that explain terrorism, retrieved from http://www.vox.com/2015/12/15/10133138/terrorism-maps, December 2015.*

[2] *Z Beaucham,. 33 Maps that explain terrorism, retrieved from http://www.vox.com/2015/12/15/10133138/terrorism-maps, December 2015.*

In addition to these facets, terrorism continues to interfere with regions beyond the Middle East and Africa. Most notably, two series of attacks in Paris, France in December 2014 and November 2015, with a combined total of over 150 casualties, have renewed the sense of urgency for Europe. Some of the perpetrators of these attacks had travelled to Syria and trained with terrorist forces, some were local to Middle Eastern countries and others were European nationals. Affiliations and time spent in Belgium were also central to the attack's unfolding. This complex breakdown of locations and participants dis- plays a variety of backgrounds, indicating an increased element of diversity among those adhering to terrorist agendas, as well as where and how they come into play.

On the 2nd of December 2015, an attack in San Bernadino, California was carried out by Syed Farook, an American citizen, and his Pakistani wife Tashfeen Malik. This combined local and international union conveys the unrelenting spread of terrorist ideology. Their differing places of birth and their travels display the international setting terrorism encapsulates. The fact that these employed, college-educated extremists did not have criminal records or personal affiliations with terrorist organizations also indicates the changing face of the terrorist profile. Furthermore, the internet's role in the story of the attack – the killers met online and Farook financed his operations using bank loans easily obtained online without sufficient screening – speaks to the current state of terrorist communications and techniques. (no author, 2015 San Bernadino Attack", retrieved from https://en.wikipedia.org/wiki/2015_San_Bernardino_attack).

Additionally, the international pull of the terrorist cause is a present factor within this discourse. Radicalization has led to defection by citizens of Western countries to the Middle East to support jihad. Figure 3 displays the multinational scope of conflict within the specified and limited region of Syria; what we learn from indications such as this is that the reach and potential for communication of terrorists is vast in nature. Further research will indicate how the role of online interconnectivity has instigated this unique and unprecedented situation.

1.2 Root Causes and Ideological Doctrine

While it is not necessarily reasonable to expect to establish a principal root cause of the terrorist mechanism, following the thread of how terrorists communicate, exchange ideas and information, and ultimately carry out acts of terror leads the investigator

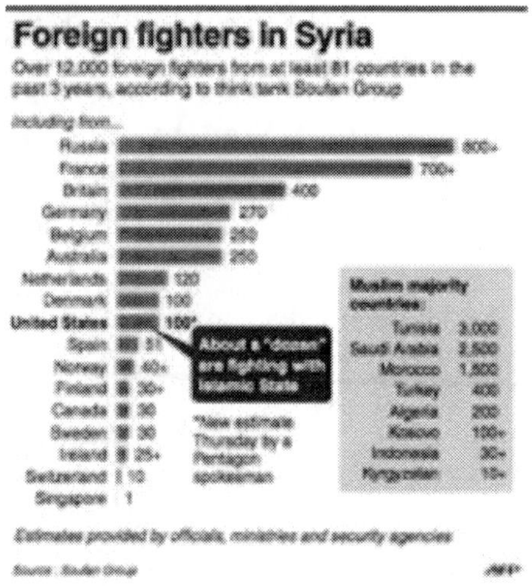

Figure 3. Foreign Fighters in Syria

to encounter numerous trends. In a *Policy Innovation Memorandum* article entitled 'A Global Venture to Counter Violent Extremism' author Ed Husain attempts to set the scene how core beliefs central to practitioners of terrorism. He writes, *'Al-Qaeda' and its ideological affiliates do not operate in a vacuum; rather, they feed off of ideas that have proliferated in Muslim communities over decades. A combination of religious literalism and conspiracist politics is at the core of their anti-Western ideology.'* (Husain 2013) Husain is not alone in rooting his perspective with the idea that terrorism's essence is ideology- based. In a *Journal of Islamic Studies* review of George Joffee's book *Islamist Radicalization in Europe and the Middle East* Humayun Ansari describes British Prime Minister David Cameron's outlook. Ansari summarizes: *'The root cause of the threat we face,' he states, 'is the extremist ideology.' It is this 'evil' – 'poisonous' – ideology that it is to blame for Muslim radicalization. He dismissed the view that historic injustices, recent wars, poverty or hardship, 'unelected leaders across the Middle East' as major contributory factors for radicalization, describing such arguments as 'grievance justification.'* (Ansari 2015, p. 116) While Cameron has positioned ideology as the central factor in extremism, Ansari refutes that it is one of several that influence terrorism, which is at least pushed into existence by other characteristics. First, he feels that Cameron's belief *'neglects what much of the research suggests, that there is no single cause but a complex of internal and external factors, catalysts, which can lead to the radicalization of individuals and groups.'* (Ansari 2015, p. 116) Of these potential other causes that aid in the process of radicalization, Ansari offers the examples of *'poor governance (repressive autocracies, exclusion of popular participation, and exploitation by corrupt and venal elites), ideological challenge, [and] economic failure,' presented as 'drivers.'* (Ansari 2015, p. 116) Thus, it can be seen that ideological doctrine is an anchor among various interconnected circumstances or ideas that formulate the basis for radicalization. It is an idea used as needed, personally relevant to varying degrees among those in the widened web of terroristic activity.

Husain's writing proposes a clearer definition of the ideological beliefs at hand. His research shows how ideology leads to specific behavior and activities organized in opposition to Western ideals. His qualification of religious literalism is at the foundation of this ideological approach. He writes that, *'These ideas include the beliefs that democracy is man-made and only extremist understandings of God's law should be enforced; that violent jihad is a Muslim obligation until 'God's law' is manifest; that those who die pursuing it, including suicide bombers, are martyrs; and that the greatest obstacle to Islam's dominance is the modern West, led by the United States. Killing Americans, therefore, weakens an enemy that oppresses Muslims.'* (Husain 2013) The tenets of such ideology are problematic in every way, and these ideas have been used time and time again to justify terror. The author concludes, *'Unless such ideas are challenged and discredited, extremist groups will continue to regenerate no matter how many terrorists are killed.'* (Husain 2013)

Also contesting this form of ideological doctrine is another person who once held the same position as David Cameron: Tony Blair. The former British Prime Minister famous for leading at the time of Britain's violent post-9/11 Middle East presence also says that the root cause of ideology is an elemental issue, not a symptom. Speaking in Paris on the 4th of December 2015, he stated, *'Defeating Daesh is only a necessary beginning'. Force alone will not prevail. The challenge goes far wider and deeper than the atrocities of the jihadist fanatics. The Islamist ideology has also to be confronted. Today it can be, in alliance with the modernizing and sensible voices within Islam determined to take the name and reputation of the faith of Islam back from the extremists. A continued failure to recognize the scale of the challenge and to construct the means necessary to meet it, will result in terrorist attacks potentially worse than those in Paris, producing a backlash which then stigmatizes the majority of decent, law-abiding Muslims and puts the very alliance so necessary at risk, creating a further cycle of chaos and violence.'* (Watt 2015) While acknowledging that radicalism hurts the everyday law-abiding Muslim, Blair is nevertheless resolute in his condemnation of the jihadist mentality's ideological output. We understand that the threats of terrorism are not limited to violent attacks, but to the spread of misguided and hateful perceptions as well.

All of these concerns help us in our understanding of how leaders react to the presence and spread of jihadist ideology. Leaders from Western and Middle Eastern countries alike have spoken out against this dangerous worldview, vowing that we must band together to halt the growth of this problematic outlook.

2. YOUTH DEMOGRAPHIC AND TECHNOLOGY

2.1 Rise of Technology and Internet Communication within the Radicalization Experience

Within recent history, school-aged children have partaken in a varied number of groups that promote or carry out acts that often lead to murder and destruction. (Johnston 2007) These youth have helped to fill the ranks of radicalized guerrilla-type organizations, militias, gangs, and terrorist groups. Their roles within these enclaves have varied, from exacting logistical information support, serving as 'lookouts' or 'mules', raising funds, taking part in battles, or carrying out attacks. The process by which young persons get involved in these groups is manifold as well, with some inheriting the role through having been brought up within radical environments that use and condone violence, some being scouted out and selected by organizations, some joining the group by personal choice, and others, for one reason or another, being forced into membership. While, of late, mostly focusing our attention on the issue of 'Child Soldiers' in rebel armies, there is increasing evidence that young people are being drawn into membership of a plethora of terrorist organizations operating internationally, including ISIS.

Proceeding concurrently to these recruitment processes was the spread of Internet usage among the average person. As online connection through email, social networking, text messaging and link sharing has become more commonplace, the immediacy of access to dissemination of terrorist material has increased exponentially. The constant connectivity and digital erosion of time zones has helped international communication, creating a virtual space within which terrorism can flourish. Additionally, increased bandwidth and video capabilities have led to more sophisticated types of information to being shared. Finally, advances in encryption technology have helped terrorists operate in secrecy.

Internet usage among youths has experienced a noticeable gain, evolving from a passive, individually directed, information seeking process to a responsive, socially networked, user-focused architecture where youth interact, discuss, dissent, create, and share content. The significance of this shift of environments is sensed when looking into the dramatic growth and popularity among youth of smartphones and with them, social networking sites such as Facebook, Twitter and Instagram, and is further exemplified by interactive video streaming websites such as YouTube. Terrorists have manipulated the size of these networks' respective user bases to spread their ideology.

The growth of social media supporting terrorist ideology has blossomed commensurately to the development and ubiquity of social media in the lives of young people. Not only does ISIS maintain accounts on popular social networking and media sites in addition to publishing a magazine, ISIS has also endeavored in developing a game software through the Google Play store (no longer available) and uses hash tags on Twitter much like a Western company might. It is in effect a brand, aware of the implication of advertising and representing itself, operating to attract, attain and retain a supportive and engaged population base.

Put succinctly, the process of terrorist organization has been greatly helped by the emergence, integration and application of these new technologies. No organization has worked these fresh trends into their operational procedures as adeptly as ISIS. Known for their consistent content generation in the form of photos, videos and press releases, ISIS has assumed a brazen approach to public relations, which has attracted international scorn, as well as new devotees.

Figure 4. Displays the Measured and Consistent Approach of ISIS Output

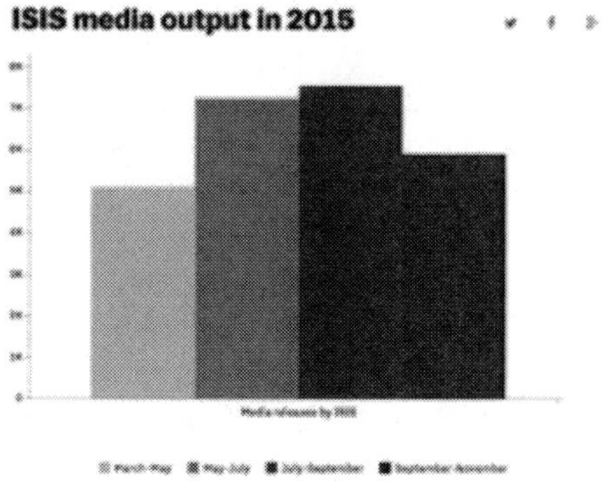

Though declining, ISIS continues to releases thousands of communiqués on a regular basis.

2.2 Who Is Susceptible?

For the most part, these young people find themselves in transitional phases of their lives: students, immigrants, people between jobs, those who have left or are about to leave their native family and in search for a new family of friends and kindred spirits with whom they can find meaning and be- longing. In this regard, we witness how ideology is decidedly not the only determining factor at play when it comes to influencing youth towards radicalization. This notion is expressed by Dr. Alan P. Schmid in *'Radicalization, De-Radicalization, Counter-Radicalization: A Conceptual Discussion and Literature Review.'* He writes, *'The importance of social networks and enabling environments has now been widely acknowledged. Social and kinship relationships are crucial in drawing vulnerable youths into a terrorist group. The process appears to be similar to the one we can also see in street gangs and religious cults. Radicalization often follows recruitment into such groups rather than preceding it. A recent US Bipartisan Policy Report noted: 'Also important are social and group dynamics, given that radicalization often happens in 'dense, small networks of friends', and that extremist ideas are more likely to resonate if they are articulated by a credible or charismatic leader'... The existence of radical milieus, whether in a concrete neighborhood or in the virtual social space of the Inter- net can be such a point of attraction for vulnerable young people in search of comradeship, a new role, identity and status, especially when push factors like discrimination, marginalization and humiliation experienced in the family, school, neighborhood and society contribute to a break with a past that is experienced as intolerable.'* (Schmid 2013, pp. 26-7) This quotation speaks to the intensity of radicalization and the delicate state of those targeted parties. Terrorists are aware of these weaknesses, and use them to their advantage, relying on the poverty, desperation and sociopolitical shortcomings of troubled regions to locate and attract new members. It is devastating but true that we can detect how real world trauma and experiences make these dislocated youths more vulnerable to what ISIS has to offer them as a form of redemption from their unfortunate situations.

This has been true in Tunisia, where a spring attack made news around the world. In *'Tunisian youth counter radicalization with innovation'* Christine Petre writes that, *'The most common explanations as to why, following the revolution, many young Tunisians appear susceptible to radicalization, include the dire economic situation, with an estimated 30% youth unemployment, poor quality of education, the failure to instill critical thinking and an overall disillusionment that has left many with little hope for the future.'* (Petre 2015) The desperate situations of the population of people who are open- minded to what terrorism can offer are the sad prerequisites for participation in brutality. Inevitably, technology has played a role in this attraction process.

3. PREVENTION AND CONCLUSION

3.1 What Is Being Done?

Despite how easy it is to become discouraged about our state of affairs, we must remain optimistic in the face of danger. Without resolve, the terrorists win. Thankfully, practices are being put in place to minimize the damaging effects of terrorism today.

Tunisia's spring attack galvanized the country, whose relative state of political disarray and geographical location makes it susceptible to the impact of ISIS. Eric Reidy reports that in Tunisia, *'the government is now emphasizing its efforts to increase security and counter radicalization. So far, security operations against militant cells and the introduction of a new anti-terrorism bill in parliament have been the centerpieces of this effort. In addition, the Ministry of Religious Affairs has been working to bring back under its control those mosques that harbored extremist imams and networks after the revolution. Part of its effort involves reinstating ministry-sanctioned leaders in these mosques. While these moves represent initial steps, parliament Speaker Mohamed Ennacuer, called for a national strategy to address the root causes of radicalization in the country. To supplement the steps that have already been taken, the government is planning to establish an anti-terrorism commission to devise a comprehensive strategy for preventing violent extremism and radicalization.'* (Reidy 2015)

Furthermore, the European Union has addressed the pressing issue in the midst of its refugee and terrorism crises. A ten million Euro pledge to invest in anti-radicalization programs has been described as follows: *'Under the new program 'Countering radicalization and Foreign Terrorist Fighters', the EU will allocate a first tranche of €5 million to fund technical assistance to enhance the capacities of criminal justice officials to investigate, prosecute and adjudicate cases of foreign fighters or would-be foreign fighters. The second tranche of 5 million euro will finance countering radicalization programs in the Sahel and Maghreb region. It will offer the possibility to non-state actors to implement activities in the field of media, education, religion, culture. Projects focusing on messaging, inter- net, social media, disengagement, and awareness-raising with social workers will also be considered as well as measures preventing the radicalization and recruitment of foreign fighters.'* (European Commission Press Release, April 2015)

The above quote's understanding of the importance of maintaining an online presence reflects the sensitivity of Internet communication within this conversation. This sentiment is echoed in Omar Ashour's article entitled 'Online De-Radicalization? Countering Violent Extremist Narratives: Message, Messenger and Media Strategy.' He writes: *'There has been a debate on the role of the Internet, whether primary or secondary, in promoting and publicizing extremist narratives, in facilitating radicalization and recruitment processes, as well as in reaching new audiences. Given the scope and the intensity of the problem, using the Internet and other media outlets to revert that role, and to counter-violent extremist narratives becomes a global imperative. As opposed to its effects on radicalization, the Internet can play a vital role in promoting a counter-narrative and in facilitating counter-radicalization and de-radicalization efforts.'* (Ashour 2010) It is interesting to note that the very same measures used to aggress can be used to heal; the power and immediacy of social media is such that it must be wielded intelligently in order to assert a presence in the lives of those young people who are at risk of radicalization.

3.2 Concrete Solutions

The recommendations included within Husain's articles are worth sharing here as well-defined and eloquently expressed examples of what can be done to limit the impact of Internet-derived radicalization among today's youth. They appear as follows:

1-Educating Muslim thought leaders in mosques and on university campuses through workshops and testimonies from former radicals about why Islamist hard-liners threaten Muslim communities. In 2009, al-Qaeda and its affiliates killed more Muslims than non-Muslims. Muslims need to reclaim their faith because Islamist extremism endangers the very fabric of mainstream moderate Islam.

2-Providing financial support to moderates for establishing alternative satellite television channels across the Middle East. A region with bulging youth populations, high unemployment, low rates of reading books, and mass popularity of satellite television means radical clerics often provide unchallenged guidance on questions of religion and politics.

3-Supporting the publication and dissemination of reading materials on normative Islam not the radical materials of Islam and targeting distribution on campuses, in bookshops, and in mosques.

4- Awareness programs targeting youth under the supervision of government agencies. Saudi Arabia has launched a national program (Faten) to protect the security of society and the social, cultural, health and economic threats. Targeting educational communities in cooperation with relevant institutions (Ministry of Social Affairs and the Ministry of Interior and Ministry of Health and Ministry of Islamic Affairs and the General Presidency of Youth Welfare and universities (sociology, psychology, social work, research chairs specialized)).The program seeks to achieve high-end goals through the preparation and design of the output, which will target communities (training packages instructors training programs awareness counseling centers campaign) and build effective relationships with relevant institutions and teams volunteer to communicate with everyone through the website and for other means of communication investors all sources of moral and material sports through partnerships with the private sector and get endowments to secure the sustainability of the program. Another youth initiative sponsored by the U.S. Department of State is global university initiative to counter violent extremism. Engaging millennial students in international campaigns to educate their peers on challenging this messaging and empowering each other through the development of a social or digital initiative, product or tool to counter violent extremism.

5-Initiating the around-the-clock presence of a professional, well-informed network of web-savvy Muslims who are active in Arab and Muslim chat rooms and on social media, refuting al-Qaeda propaganda with factual and scriptural arguments. The purpose is not to dissuade jihadis (that would be a bonus), but to ensure virtual audiences do not assume that extreme narratives are unchallenged and hence preponderant. (Husain 2013)

These nuanced and detailed ideas are customized to the unique needs of an increasingly inter- connected society. We can learn a lot from what is wrong in order to push what is right. Husain's suggestions are in tune with the need to zero in on exactly what must be done consistently to stop the spread of terrorism. These are the sorts of ideas that directly confront the challenges listed in the re- search consulted for this paper. With great hope, they are presented here as evidence that all is not lost in the war on terror.

REFERENCES

- BBC News (2015). What Is 'Islamic State'? BBC News. N.p., 2 Dec. 2015. Web. 27 Dec. 2015. http://www.bbc.com/news/world-middle-east-29052144 .
- Wikipedia (2015). Attack on San Bernadino.". Wikimedia Foundation, n.d. Web. 27 Dec. 2015. https://en.wikipedia.org/wiki/2015_San_Bernardino_attack#Syed_Rizwan_Farook.
- European Commission (2015). EU Provides €10 Million to Counter Radicalisation in the Sahel-Maghreb and Stem the Flow of Foreign Fighters from North Africa, the Middle East and Western Balkans.". N.p., 28 Apr. 2015. Web. 15 Dec. 2015 <http://europa.eu/rapid/press-release_IP-15-4866_en.htm>.
- Humayun, A. Islamist Radicalization in Europe and the Middle East. Edited by George
- Joffe (2016). *Journal of Islamic Studies, 27(1),* 116-118.
- Ashour, O. (2010). Online De-Radicalization? Countering Violent Extremist Narratives: Message, Messenger and Media Strategy | Ashour | Perspectives on Terrorism. Perspectives on Terrorism 4.6 (2010): n. pag. Web. 22 Dec. 2015 <http://www.terrorismanalysts.com/pt/ index.php/pot/article/view/128/html>.
- Beauchamp, Z. (2015). 33 Maps That Explain Terrorism. Vox. N.p., 15 Dec. 2015. Web. 20 Dec. 2015 <http://www.vox.com/2015/12/15/10133138/terrorism-maps>.
- Beauchamp, Z. (2015). ISIS's Online Propaganda Machine, in One Chart." Vox. N.p., 07 Dec. 2015. Web. 20 Dec. 2015. <http://www.vox.com/2015/12/7/9867718/isis-online-propaganda-chart>.
- Husain, Ed (2013). A Global Venture to Counter Violent Extremism. POLICY INNOVATION MEMORANDUM NO. 37 Date: September 9, 2013 37 (n.d.): n. pag. Council on Foreign Re- lations. Web. <http://www.cfr.org/radicalization-and-extremism/global-venture-counter-vio- lent-extremism/p30494>.
- Johnston, Ph. (2015). MI5: Al-Qaeda Recruiting U.K. Children for Terror. Telegraph.co.uk, 7 No- vember 2007. Web. December 23, 2015.
- Petre, Ch. (2015). Tunisian Youth Counter Radicalization with Innovation. Voices and Views: Middle East and North Africa. World Bank, 31 July

2015. Web. 22 Dec. 2015. <http:// blogs.worldbank.org/arabvoices/tunisian-youth-counter-radicalization-innovation>.

- Reidy, E. (2015). Tunisia Cracks down on Radicalization - Al-Monitor: The Pulse of the Middle East. Al-Monitor. N.p., 19 May 2015. Web. 21 Dec. 2015. <http://www.al-monitor.com/pulse/originals/2015/05/tunisia-government-counter-extremism-anti-terrorist-bill.html>.
- Schmid, A. (2013). Radicalisation, De-Radicalisation, Counter-Radicalisation: A Conceptual Dis- cussion and Literature Review. Terrorism and Counter-Terrorism Studies ICCT Research Papers, 1-61.
- Watt, N. (2015). Support for Isis Stretches 'deep into Parts of Muslim Societies,' Says Blair. The Guardian. N.p., 4 Dec. 2015. Web. <http://www.theguardian.com/politics/2015/dec/04/ support-isis-parts-muslim-societies-tony-blair>.

Global Issue: Shortage of Male Primary Teachers

Yang WU
Higher Degree by Research (HDR) Student, Deakin University, Australia

1. INTRODUCTION

Gender issue in education is a hot topic discussed for several years. Education plays a significant role in ensuring that males and females have the same opportunities in their personal and professional development, through formal schooling, shaping attitudes and transforming their behaviors. Specifically, primary level education is fundamental for building up students' worldview, which has far-reaching influence on their future life. Instead of emphasizing gender equity, gender equality is widely concerned by different nations all over the world. Gender equality does not mean that men and women should become the same but that a person's rights, responsibility and opportunities should not depend on whether they are born male or female (UN Women 2014).

Current report will shed light on the gender distribution among primary teachers – shortage of male primary teachers. From statistical data, we can see that it is not an issue within certain areas but a cross-national and cross-cultural issue. OECD report shows that, on average in OECD countries, the majority of teachers are females. Specifically, 82% of teachers at the primary level, 68% at the lower secondary level, and 56% at the upper secondary level are female (OECD, 2014a). The phenomenon that female teachers play a dominant role in lower level teaching positions, is widely found in both Western and Asian countries. Thus, this paper will try to find out the similar values shared by people from different societal cultures and traditions, in terms of shortage of male primary teachers.

Through reviewing official statistical data and reports, the condition of lacking male primary teachers in Western and Asian countries will be briefly presented. Then, the significance of male primary teachers will be explicated, followed with discussion of possible reasons explaining such shortage. Specifically, lower salary and social stereotype of primary teaching profession, as two main factors will be discussed. In the last part, some suggestions will be provided to reflect this global issue. Qualitative equality is more emphasized than quantitative/statistical equality referring to gender distribution among primary teachers in this report.

2. SHORTAGE OF MALE PRIMARY TEACHERS IN WESTERN COUNTRIES

Shortage of male primary teachers widely exists in most Western countries. The American statistics shows that, around 20 years before 2005, only 21% of public school teachers were male. In lower grades, the proportion of male and female teachers tended to be more imbalanced – male primary teachers only occupied 9% of total populations. Some American public primary schools even have not had any male teachers for 12 years (Liu, 2005). Moreover, in 2007, the official data shows that the number of American male teachers came to the lowest point, and it is much severer in primary level (Xinhua Daily News, 2007). In 2008, it was also found in Ontario (a province in Canada), the proportion of male primary teachers is less than 10%, and this figure still keeps decreasing (Parr, Gosse & Allison 2008). Australian labor market report revealed that, recently shortage of primary male teachers widely exists in local public, private and religious schools. Additionally, this issue even appears in some secondary and high schools (Chinese Teachers Newspaper 2006). Furthermore, the data from New Zealand's Ministry of Education (1996 & 2005) disclosed that, through 50 years, this country's primary teachers dropped from 42% to 18%, and this figure is still declining.

Compared with above countries, European data tends to be more acceptable – approximately 15% male primary teachers on average. In 2012, *English Daily Mail* published an article disclosing the situation of English primary teachers. It said that one out of five boys did not have any class taught by male teachers at primary level. That is to say, around 360485 boys aging 4 to 11 only attend class taught by female teachers. In Scotland, male primary teachers only contribute 7% of total teacher populations (MacKinnon, 2003). OECD report (2014b) also showed that the vast majority of primary teachers are females across the OECD – approximately 8 out of 10. In other words, around 20% primary teachers are male. Therefore, early in 2005, the international conference hold by OECD in Paris pointed out that, shortage of male teaching professions, which is considered as a long-term and cross-national trend, does need more attention.

Over several years, education policy makers from different countries (e.g. America, England, Australia, Canada, Finland, New Zealand, Scotland and etc.) showed high concern about male entering education domain, especially primary level education. Through reviewing official report and literatures, it is found that shortage of male primary teachers widely exists in the majority of Western countries as well as Asian countries.

3. SHORTAGE OF MALE PRIMARY TEACHERS IN ASIAN COUNTRIES

Generally, the ratio of male teachers in primary schools in Asian countries is relatively lower than the ratio of female. In China, the ratio of male primary teachers is around 20%. In some areas, male primary teachers are less than 10% of whole teachers. *Chinese Primary and Secondary Teachers Development Report* (Minister of Education 2012) pointed out that, China has experienced feminization of primary and secondary teachers for over 10 years. More than 80% of primary teachers in urban areas are females. The number of newly recruiting male teachers is declining, and in-service male primary teachers are 'escaping' which leads to less male teachers in primary schools in the future. Some Chinese educators addressed lack of male primary teachers is a significant issue needing more attention.

The proportion of male primary teachers in other Asian countries is also lower than female counterparts. One comparison study exploring gender distribution among school teachers in Taiwan, Japan and South Korea (Huang, Yang & Wu 2012) found that, female teachers are the main human resources in primary schools in these three areas. Specifically, the proportion of female teachers in Japanese primary schools experienced a moderate decrease from 65% in 2002 to 62.7% in 2007. That is to say, the proportion of male primary teachers increases during this period, but still less than 40%. In contrary, the proportion of primary female teachers in South Korea increased from 70.3% in 2000 to 76% in 2006. In other words, the ratio of male primary teachers decreased to approximately 24%. Moreover, it is also revealed that the number of newly recruiting male primary teachers only occupied 21.7% of total teacher populations in Korea (Chung 2006).

However, in India, the situation of male and female primary teachers is opposite to above Asian countries. Until recently, teaching in India has been still a male domain, even in primary schools (Chudgar & Sankar 2008). In the middle-1970s, there was only around 25% female primary teacher labor force (Seetharamu 2002). In order to hire 50% female teachers, the intervening program named Operation Blackboard is initiated since 1987. Recent data showed that female teachers contribute up to 43% but still less than 50% (Department of Elementary Education and Literacy and Department of Secondary and Higher Education 2004).

From the review of official statistical data and reports, with some exceptions (e.g. India), the reality of less male teachers in primary grades continues to be a topic of debate in most countries all over the world. Gender imbalanced environment within primary school teaching teams will potentially influence students learning knowledge, constructing sexual roles, as well as building up their worldview. The reasons explaining the shortage of male primary teachers in most countries should be explicated. Some intervening programs are in need to attract more males joining primary teaching teams as well as to keep in-service male primary teachers continuing their teaching profession. In following parts, these issues will be emphasized.

4. THE SIGNIFICANCE OF MALE PRIMARY TEACHERS

Gender-balanced distribution of primary teachers is beneficial for both male and female students' development in various aspects – capacity of knowledge learning, inter-relationship and construction of gender identity. A male model is also irreplaceable in children's young years. However, feminization of primary school, on the other hand, refers to shortage of male teachers in primary school statistically, culturally and with certain backlash of feminist movement. The following paragraph, I will discuss each of them in more details.

From statistical perspective, feminization in primary school means that females make up the majority of teaching forces, which makes school predominantly staffed by female teachers (Harnett & Lee 2003). Statistical data showed that more 70% of primary teachers are female in most countries. Culturally, because of dominantly female roles, it inevitably makes environment of primary schools feminine (Skelton 2012). These feminine ethos and cultures in primary schools could potentially tend to favor girls and ignore boys; even discriminate against male staffs and students in such situation. Furthermore, some researchers (e.g. Skelton 2012) presented that, the criticism and accusations of feminist movement, leads to gender inequality in education, especially in primary teaching professions. Because of different characters and teaching style based on genders, male and female primary teachers have their own advantages to be good for children' development. Feminization of primary school – shortage of male teachers, could 'damage' students (especially boys) and teaching teams in following ways.

Compared with girls, boys could be more struggling in primary schools, because of lacking male teachers. Some researchers (e.g. Skelton 2002) suggested that, the feminization environment in primary school and female-dominant pedagogy of schooling position girls to be the 'winner', while boys to be the 'victims'. In other words, female-dominant teaching environment makes primary school potentially favor girls' learning style over those of boys. In order to provide primary students (specifically boys) with male role models, it is necessary to encourage males to enter a caring profession (DeCorse & Vogtle 1997). Substantial longitudinal studies are conducted to explore the impacts on students' academic achievement when having more male teachers. The findings are impressive – little or no impacts are found on students' academic achievement when having more male teachers (Johnston 2005). However, it is significant to note that since these studies are conducted in a short-term, it does not mean there is little or no far-reaching influence on students' lives and their future development when having more male teachers.

Moreover, Students' capacity of knowledge learning, inter-relationship and construction of gender identity could be influenced in different extent, because of lacking male teachers. Feminine teaching environment is not beneficial for students' knowledge learning and development of inter-relation with others, because this gender-imbalanced environment cannot provide gender-equal models for both boys and girls. Especially, for those students (no matter boys or girls) who do not have many positive male models in their lives, male primary teachers can serve as models and contribute to students' development of positive gender identities (OECD 2015).

More significantly, shortage of male primary teachers currently can impact next generation's understanding of primary teachers, and potentially influence future recruitment of male teaching professions in primary schools. Let's imagine that, if most primary teachers are female, and very few male teachers mainly teaching math-related subjects and sports, what cultural messages will boys and girls receive in this experience? How will it influence their future learning and adult life? It could build up students' gender stereotype in these areas. They will naturally consider teaching profession, especially working with young children, is female's job, and male will not be suitable to enter into this field – less male primary teachers in the future.

From teachers' perspective, the efficiency of teamwork can be enhanced under gender-balanced working situation. Gender segregation in career choosing leads to talent loss for both individual and society. Some researchers confirmed that significance of gender diversity in a team. For example, Hoogendoorn et al. (2011) suggested that gender-diverse team may have greater success, and Woolley et al. (2010) also mentioned about better problem solving can be attributed to having more females in a group.
To Sum up, it is important to have some males in primary teaching profession, as shortage of male primary teachers is unbeneficial for students' development, especially boys, and primary teaching groups. Specifically, I mean 'some' instead of 'as equal number as female teachers'. Because of cultural and societal issue, it is unrealistic to achieve exactly the same number of male and female teachers in primary schools.

5. THE POSSIBLE REASONS EXPLAINING THE SHORTAGE OF MALE PRIMARY TEACHERS

The significance of primary teaching forces that is representative of both sexes is a goal worth aiming for. In order to develop proper strategies to achieve it, we need more understanding of underpinning reasons that why less males get involved in primary teaching. Through reviewing journal articles and international reports, there are numerous reasons explaining the shortage of male primary teachers. In current report, two of them are emphasized as main factors – salary and social stereotype.

5.1 Salary

Salary is at the top of list to explain the lower numbers of male primary teachers. Cushman (2007) summarized that teaching salaries became a major concern for those considering a career in primary teaching, and for those who have already ensconced in teaching profession. Study conducted by Parr, Gosse, and Allison (2008) exploring the perception and experience of male teacher candidates, addressed the confliction of role demands. Male teacher candidates involving in this study felt that it is their responsibility to hold a job and financially contribute for their family. In other words, male encountered with multiple social roles, and the dominant one is breadwinner that is shared in many cultures.

Moreover, males concern more about salary than female counterparts. Thornton and Bricheno (2006) discovered that males tend to be more responsive to financial incentives. It is also pointed out by Parr et al. (2008) that, giving up a secure salary (maybe not very high income) and returning to school to obtain a teaching certificate (it is required to become a teacher) that might lead to a teaching job, is a great gamble for many males. Low salary may make the teaching professions lose attraction for potential males teachers. In order to attract more males, raising primary teachers' salary is included in some reports (e.g. OECD 2004) as one of main measures to moderate this issue. However, it is not enough. More understanding of social stereotype about male primary teachers could provide further views.

5.2 Social Stereotype

5.2.1 social status

In the OECD country reports, the status of primary teaching, similar in other educational publication, continues to be regarded as low-down on the list of middle-class profession. Even in the nineteenth and early twentieth century, teaching itself was not considered as a high status vocation for males (Skelton 2012). Skelton mentioned that some antifeminists' voice that lower status of primary teaching is attributed to absence of males. Moreover, the association of caring and nurturing with primary teaching is a significant factor, which makes the impression that primary teaching profession is a low-status women's vocation (Cushman 2007). Thus, compared with other professional vocations (e.g. engineering, manager.), primary teaching is considered to be with relatively lower status in many countries.

5.2.2 as a male primary teacher

A male primary teacher could encounter some misunderstandings and embarrassing situations, as he is working in a socially considered feminized environment – primary schooling. A small-scale study of male teachers who are working with young children, provided evidence that how males themselves and female colleagues are uncomfortable at finding a male doing a 'female' job (Williams 1993). Skelton (2012, p. 8) summarized some provocative voice such as 'What does it mean when a man teaches young children? Can't he get a 'proper job'?' Because of prevailing view of primary school as feminized, males who choose to work there, to some extent, are not seen as 'real men' culturally (Skelton 2003). Some male primary teachers reflect that their ability of caring and nurturing children is doubted by female colleagues, just because they are men. Moreover, male primary teachers are regularly characterized as 'feminine', 'homosexual', and 'pedophile', both from within the profession and in the public eyes (Parr et al. 2008, p. 258). There is a strong gender stereotype underpinning it.

The issue of working with young children also put male teachers under sensitive topic. In some countries (e.g. New Zealand, Australia), fears of being accused of child abuse and paedophilia are continuing to act as the major hinder for male to consider the career as a primary teacher (Cushman 2007). In Parr and colleagues' (2008) study, male teacher candidates noted that they cannot conduct the work as teachers smoothly, because of some children protection policies. For instance, the 'no touch' policy with students reportedly interfere male teachers' natural inclination to address children's need when they perceive. Male teachers, especially male primary teachers, will be hesitant to offer or afraid of giving a hug or some in-touch body languages to encourage students when they think it is in need

5.2.3 vocational development

In many countries, males rapidly start from a primary teacher to school principal, as socially, males are more favored in the process of employment and promotion than females. The result of flourishing vocational development is less male primary teachers in the classroom (Cushman 2007). Traditional gender stereotype of primary teaching stops many males outside of primary schools' gates. On the other side, traditional male leadership stereotype promotes most in-service male primary teachers to administrative positions or leadership roles. Less and less male primary teachers are left in classrooms, which makes primary teaching team is mainly made up of females. From an outsider perspective, feminized primary schooling comes into the point.

There are also some other reasons underpinning the shortage of male primary teachers, such as 'attribution from teacher education and teaching' (Cushman 2007); 'supports and social networks' and 'previous experience' (Parr 2008) and etc. It is significant to understand why many males stop outside of primary teaching in different cultural contexts.

6. DISCUSSION AND SUGGESTIONS

In 2004, referring to 20 or more countries' reports submitted to OECD on attracting and retaining teachers, all revealed a decline in the numbers of male coming into teacher profession, especially at primary level. And some countries even note that this is a case for both male and female, and needs more concern. Much more balanced gender distribution among primary teachers has positive effects on both male and female students. Some intervening programs, cooperation and policies are initiated to change the situation. During primary teachers recruiting process, positive gender discrimination towards males is utilized to attract more men into this field (Cushman 2007). It is also suggested that the cooperation among government, schools and private sectors will be initiated to decrease gender stereotype in some careers (e.g. primary teachers) with certain gender label (OECD 2012). Furthermore, raising primary teachers' salary is written into the report (OECD 2004) to change social status of male primary teachers.

Besides, recruiting more male primary teachers to achieve equal numbers of male and female teachers in school, it is more significant for us to understand this global issue neutrally and comprehensively. In other words, qualitative equality is more emphasized than quantitative/statistical equality in this report. Historically, we used to have more males in teaching profession, even for primary teaching. For example, in the ancient China, the word 'teacher' is only for male. Several decades ago, teacher is considered as a male social role in western cultures. Because of changes of economy and development of society over years, structure of society changed including the social role of teacher. It gradually shifts from a male domain to a female domain in many nations.

Gender imbalance among primary teachers is not just an issue within education. It is a broader topic involving economy, policy and situation of world development. The balance point is at the cross of social supply and demand. The new policies and intervening programs are to stimulate the supply or demand, and then it will achieve the new balance temporarily – the new gender distribution of primary teachers. More importantly, we need to be aware that equal number of male and female teachers in primary school is not always good for the development of students, primary schools or society. The social and cultural contexts can never be ignored. The balanced situation is never statistically equal. We need to have more understanding about male primary teachers' perception of and attitudes towards their own jobs, and how social cultures and traditions are changed with time, and how do these changes make a difference of male primary teacher from before.

Although, over last decades, great advances have been made, we still do not live in a world where men and women have the equal opportunities. Gender inequality exists in schools, universities and more broadly in society. The shortage of male primary teachers is only part of this issue. We need to have patience. Anything relevant with gender, culture, stereotype and worldview, can never be solved in a short time or within one generation. It takes several generations to make even tiny but permanent changes. How can people who lived hundreds years ago imagine that teacher – a male dominant career – will be feminized in this day and age?

REFERENCES

- Chinese Teachers Newspaper (2006). Low salary stopping male to become teachers. 4(12), A04.
- Chudgar, A. and Sankar, V. (2008). The relationship between teacher gender and student achievement: evidence from five Indian states. *Compare, 38(5),* 627-642.
- Chung, A. (2006). Korea faces shortage of male teachers. Retrieved March 3, 2007, from the World Wide Web: http://times.hankooki.com/lpage/nation/200602/ kt2006021417503011960.htm.
- Cushman, P. (2007). The male teacher shortage: A synthesis of research and worldwide strategies for addressing the shortage. *KEDI Journal of Educational Policy, 4(1).*
- DeCorse, C. J. B. and Vogtle, S. E. (1997). In a complex voice: The contradictions of male elementary teachers' career choice and professional identity. *Journal of Teacher Education, 48(1),* 37-46.
- Department of Elementary Education and Literacy and Department of Secondary and Higher Education (2004). Annual Report 2004–2005. New Delhi: Ministry of Human Resource Development, Government of India.
- Harnett, P. and Lee, J. (2003). Where have all the men gone? Have primary schools really been feminised? *Journal of Educational Administration and History, 35(2),* 77–86.
- Hoogendoorn, S., Oosterbeek, H., and van Praag, M. (2011). The Impact of Gender Diversity on the Performance of Business Teams: Evidence from a Field Experiment, Tinbergen Institute Discussion Paper. Amsterdam School of Economics, University of Amsterdam.
- Huang, Y.-C., Yang, C.-C., and Wu, H.-H. (2012). The comparison of gender distribution among

school principals and teachers in taiwan, japan, and south korea. *International Education Studies, 5(4),* 1.
- Johnston, C. (2005). Male teachers don't benefit boys, study finds (ETFO). Guardian Unlimited, September 9, 2005.
- Liu, J. (2005). American primary and secondary schools: Calling for male teachers. *Chinese Societal Newspaper, 8(1),* 007.
- MacKinnon, S. (2003, September 10). Call to boost male teacher numbers in Scots schools. *The Scotsman, 4.*
- Minister of Education (2012). Chinese Primary and Secondary Teachers Development Report (2012). Press: Education and Science.
- Ministry of Education (1996). Women in the teaching service. Education Trends Report, 8(1), 1-6.
- Ministry of Education (2005). Statistics on teaching staff. Retrieved July 11, 2005, from the World Wide Web: http://www.minedu.govt.nz/index.
- Organisation for Economic Co-operation and Development (OECD) (2014b). PISA 2012 Results: What Students Know and Can Do (Volume I, Revised edition, February 2014). Student Performance in Mathematics, Reading and Science, PISA, OECD Publishing, Paris.
- Organisation for Economic Co-operation and Development (OECD) (2004). Attracting and Retaining Effective Teachers – Country Background Reports.
- Organisation for Economic Co-operation and Development (OECD) (2015). Trends Shaping Education 2015 Spotlight 7, OECD Publishing.
- Organisation for Economic Co-operation and Development (OECD) (March, 2014a). Education Indicators in Focus: education data, education evidence, education policy, education analysis, education statistics.
- Organisation for Economic Co-operation and Development (OECD) (May, 2012). Gender Equality in Education, Employment and Entrepreneurship: Final Report to the MCM 2012. Meeting of the OECD Council at Ministerial Level.
- Parr, M., Gosse, D., and Allison, J. (2008). The professional journey of male primary teachers: Experiences and perceptions of entering into a sacred female space. *The International Journal of Diversity in Organisations, Communities and Nations, 7(6),* 257-265.
- Parr, M., Gosse, D., and Allison, J. (2008). The professional journey of male primary teachers: Experiences and perceptions of entering into a sacred female space. *The International Journal of Diversity in Organisations, Communities and Nations, 7(6),* 257-265.
- Seetharamu, A. (2002). Status of elementary teachers in India: A review. India Education Report: A Profile of Basic Education. New Delhi: NIEPA.
- Skelton, C. (2002). The'feminisation of schooling'or're-masculinising'primary education?[1]. *International Studies in Sociology of Education, 12(1),* 77-96.
- Skelton, C. (2003). Male primary teachers and perceptions of masculinity. *Educational Review, 55(2),* 195-209.
- Skelton, C. (2012). Men teachers and the 'feminised' primary school: a review of the literature. *Educational Review, 64(1),* 1-19.
- Thornton, M. and Bricheno, P. (2006). Missing men in education. Stoke on Trent, United Kingdom: Trentham Books.
- UN Women (2014). Concepts and Definitions. Retrieved from: www.un.org/womenwatch/osagi/conceptsandefinitions.htm
- Williams, C. (1993). Doing 'women's work': Men in nontraditional occupations. Newbury Park, CA: Sage.
- Woolley, A.W., Chabris, C.F., Pentland, A., Hashmi, N., and Malone, T.W. (2010). Evidence for a collective intelligence factor in the performance of human groups. *Science, 330(6004),* 686-688.
- Xinhua Daily News (2007). The shortage of American male teachers: Low salary and being blamed. 15th November: 003.

Education and Obesity

Abdullah Al-Bawardy
Student, Al-Yamamah University, Saudi Arabia

1. INTRODUCTION

Throughout the decades the amount of evolution and civilization the world has gone through is mesmerizing. This progress could not be achieved if education was not constantly encouraged. Education is our most valuable asset; it should be cherished, nurtured and fostered. As education is the base foundation of a successful society, the sense of knowledge empowers us as humans; it gives us the ability to think, create, solve and thrive. The huge impact that education had on our lives is crystal clear, from the revolution of vaccination to the creation of smartphones. The sequence of events and upgrades that we have gone through is a blessing in disguise, even though it have its perks, it also have its side effects. This article will explore the effect of obesity on cognition and education.

2. IMPORTANCE OF EDUCATION

The (identity, socioeconomic status and well-being) research, that was funded by ESRC's Secondary Data Analysis Initiative, reveals that education has become one of the clearest indicators of life outcomes in regards to employment, income and social status, and it's a strong predictor of attitudes and wellbeing. Researchers studied the association of education and wellbeing, and based on their analysis they found that people tend to improve their social identity and seek amendment, which eventually lead to a better wellbeing.

Education is one of the pillars of human's thriving. It has been one of the most powerful assets in the human self-value and empowerment of the society. Without education the human being wouldn't achieve what they achieved nowadays. In addition, education also has its print on economy, where it plays a critical role in the economical industry, where a person with a solid knowledge of the financial system and its branches such as (banking, stock exchange, currency value and monetary impact in the current economy) can positively expand the station of our country.

'The most important caveat for the literature on education and growth is that it sticks to years of schooling as its measure of education - to the neglect of qualitative differences in knowledge' (Hanushek & Wößmann 2007). Solid and concrete background of education reflects the person's self-knowledge on the way they think, what matters to them and how to use this knowledge for improvement in the field. Furthermore, it also has its value in the human skills and how to capture those skills in the self-development.

3. OBESITY

With the advancement of technology, kids became more dependent on virtual activities. This move generates a routine that revolves around sitting in front of a screen for hours on a daily basis. Unfortunately, by doing so it negatively affects their health, as exercise is no longer a part of kids' lives, they've forgotten how to go outside and actually move a muscle. In addition to that, poor nutrition and constantly eating junk food have enormously contributed in increasing the ratio of obesity and eventually leading to a poor academic performance.

4. PREVALENCE OF OBESITY

According to WHO, in 2008, 35% of adults aged 20+ were overweight (BMI ≥ 25 kg/m2) (34% men and 35% of women). The worldwide prevalence of obesity has nearly doubled between 1980 and 2008. In 2008, 10% of men and 14% of women in the world were obese (BMI ≥30 kg/m2). An estimated 205 million men and 297 million women over the age of 20 were obese – a total of more than half a billion adults worldwide. Thus, the body mass index increases with higher income, as it's shown in the Figure 1.

Figure 1

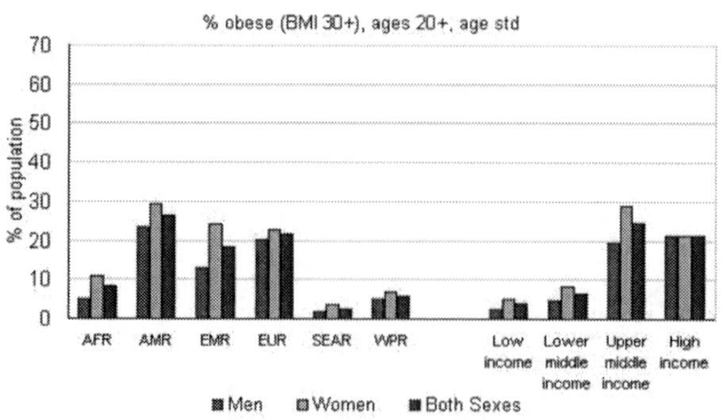

Childhood obesity is one of the most serious public health challenges of the 21st century. The problem is global and is steadily affecting many low- and middle-income countries. The prevalence has increased at an alarming rate globally. In 2013 the number of overweight children under the age of five is estimated to be over 42 million. Close to 31 million of these are living in developing countries.

5. WHAT IS THE EFFECT OF OBESITY ON EDUCATION?

Multiple researchers partaken an interest in this subject, as childhood obesity became an epidemic. Obesity is a known risk factor for multiple diseases, but the link between obesity and cognition is not yet settled. Many studies showed a positive association, the university of Illinois study explored the relationship between increased body weight and children slow reaction to stimuli, and in addition they studied the level of activities in the cerebral cortex during action monitoring. Action monitoring is when our brains are used to a certain action that we start doing it automatically without actually thinking of the mechanism of the action. For example, '*when we type we don't have to be looking at the keyboard or screen to realize that we've made a keystroke error*' said Charles Hillman (2014). In the study, they measured the behavioral and neuroelectical responses of two groups of children, one group obese and the other normal weight, using caps that recorded electroencephalographic activity while the children were asked to participate in a task presented to them. Following the test we found that children with normal weight have high capacity to evaluate their behavior and the need to change their behavior in order to avoid future errors.

Among the complications of obesity, metabolic syndrome is one of the common complications, and it has been found that metabolic syndrome impairs the cognitive function. In a study conducted by Yau, Castro, Tagani, Tsui, and Convit (2012), Adolescents were tested based on a number of different criteria, including intellectual functioning and academic achievement. The results were as follow, adolescents with metabolic syndrome scored lower in intelligence quotient (IQ), reading, spelling and arithmetic respectively. They also demonstrated a shorter attention span and decreased mental flexibility.

In addition to academic skills, research has shown links between weight status and social and emotional skills and beliefs. For instance, girls who were overweight, they also had struggles related to social and emotional behaviors (Datar & Sturm 2006). These overweight females struggled to interpret and thus communicate their feelings appropriately as well as connect with their peers. Obese adolescents also face social consequences in addition to academic ones. Moreover it's an important area to examine in its own right, ultimately, social and emotional distress and behaviors contribute to the way children learn and react in the classroom. Furthermore, overweight children tend to have low self-esteem and self-confidence, which, may end up influencing their personality shaping while growing up and as a result they will develop weak personalities not fit for leadership and management.

Few studies examined actual grades as an outcome measure when assessing academic performance and weight status. This is important, as grades tend to be more subjective in nature (Shore SM, Sachs ML & Lidicker Jr. 2008). Number of studies evaluated the direct link between grades and weight and the end results showed that obese kids have lower grade in comparison to normal weight kids. A study was piloted in Southeast Texas; eleven schools with a diverse ethnic background were included in the study. The study concluded that children in the obese category have significantly lower grades, and Children in the overweight weight category did not significantly differ from those in the normal weight category, as shown in Figure 2.

Figure 2

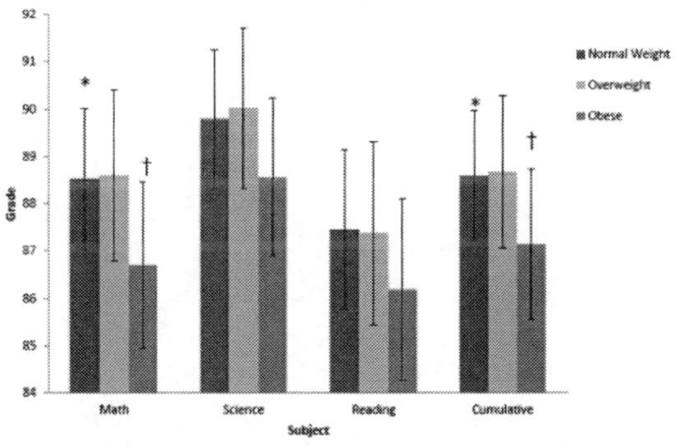

6. CONCLUSION

In conclusion, it's true that the results are somehow controversial, but obesity is a phenomenon we can't ignore, and whether it's directly related to low academic performance or not, it's strongly related to a variety of diseases, that in fact harm your health. Our children are the unending resource; they're the way to a prospering community. It's only fit to protect them and prevent any kind of harm physically and intellectually.

7. RECOMMENDATIONS

We should focus on raising awareness in regards to the importance of education and its true assets, encouraging citizens to learn more and constantly educate themselves. As there is not limit for knowledge. We can grow our knowledge in a daily basis, through reading books, watching the news, and essentially using technology. Technology can be used in different ways, you can access the Internet and educate yourself in whatever subject you want, you can also upload application that help strengthening your memory and stimulate your brain cells. There is infinite way on how you can educate and develop yourself, we only lack the will.

Obesity is a problem that can't be overlooked, our kids should be aware of the consequences of sitting for hours in front of a screen and lacking of movement, and they should know the benefits of exercise. It's our job to integrate exercises in their lives, and try to present it as fun exciting game. Exercise is mandatory, it shouldn't be stopped even if we're fit or we grew older.

REFERENCES

- Childhood overweight and obesity (2016). Retrieved from http://www.who.int/dietphysicalactivity/childhood/en/.
- Easterbrook, M., Kuppen, T., and Manstead, T. (March 2013). Identity, Socioeconomic Status and Well-Being. Retrieved from http://sites.cardiff.ac.uk/issw/.
- University of Illinois College of Agricultural, Consumer and Environmental Sciences (1 April 2014). Child's obesity, cognitive function linked, study finds. Retrieved from www.sciencedaily.com/releases/2014/04/140401131135.htm.
- Naticchioni, K. M. (Spring 2013). The Relationship between Obesity and Academic Achievement of School-Age Children. Retrieved from http://collected.jcu.edu/cgi/viewcontent.cgi?article=1008&context=honorspapers.
- Johnston,C.A., Moreno, J.P., Tzu-an Chen, Stansberry, S.A., and Woehler, D. (2014). Impact of Pediatric Obesity on Grades in Elementary School. Retrieved from http://digitalcommons.library.tmc.edu/cgi/viewcontent.cgi?article=1162&context=childrenatrisk.
- Hanushek, E.A. and Wößmann, L. (2007). Education Quality and Economic Growth. The International Bank for Reconstruction and Development, The World Bank
- Aghion, P., Boustan, L., Hoxby, C., and Vandenbussche, J. (March 2009). The Causal Impact of Education on Economic Growth: Evidence from U.S. Retrieved from www.brookings.edu/economics/bpea/bpea.aspx

Session II: Empowerment and Opportunity

Shared Governance Model as a Student Initiative Acceleration Tool ITMO University Case

V. J. Arellanes, A.K. Saitgalina, A.P. Kuropyatnik, E.Y. Tiguntseva
Master level students, ITMO University, Russian Federation

ABSTRACT

Every university has its own strategic goals. For ITMO University, the main strategic goal is to make it more competitive through improving educational and academic processes. Both students and university take part in educational process but usually it is the university who develops the educational program. In order to get a constructive feedback and to motivate students to be involved in achieving long-term goals, following shared governance concept ITMO University established the Student Office "5-100S". The main idea is to delegate some competences from the university administration to a special department, which consists of highly skilled students who are able to help others to bring their own ideas to life using project management techniques. Student Office "5-100S" is a working feedback and a mentor at the same time. Although the Student Office "5-100S" was recently created, ITMO University experience can be useful to other universities.

1. SHARED GOVERNANCE (SG) MISSION

Shared governance in a university means collaboration between university's administration, academic staff and university's students, giving them the opportunity to prove themselves implementing evidence-based practice. It means that part of everyday tasks and administrative responsibilities are entrusted to the students that are capable of managing them. So actually the shared governance concept is already used in every university but it started to develop just some time ago. Involving students in guidance of their university gives benefits to its administration: decreasing the amount of work for key managers, preparing future experienced university employees, fast feedbacks from students and new points of view and ideas about university's processes.

1.1 Student Office "5-100S"

The future of every university is formed by its students' generations; every generation brings their own new ideas and peculiar projects for solving their present-day problems; that is the reason of why involving youth in the global development process of the university is one of the key goals of the ITMO's University strategic long-term sustainable development. The most important and vital strategic goal is our mission, which is described as follows:

- A part of The SG mission is to prepare a young management team, capable to work in the conditions of dynamically developing sphere of higher education by the year 2020.

At the moment ITMO University participates in a Global National excellence program of competitiveness «5-100-2020», 5 universities getting to the top 100 World's University ranking for the 2020 year. The goal of the Project is enhancing the competitiveness of Russian universities amongst leading educational centers. For ITMO University it means to achieve a leading position among the world's educational and research elite by introducing cutting-edge research in the field of IT and Photonics, Robotics and New Materials, Biotechnologies and Health and on this basis the development of highly skilled new generation. It was decided to realize and to start up Student Office "5-100S" (fig. 1) following the roadmap of "5-100 Project" of ITMO University in order to prepare active students for management capabilities that will assist them to become university officials in future.

Figure 1. "5-100 S" office

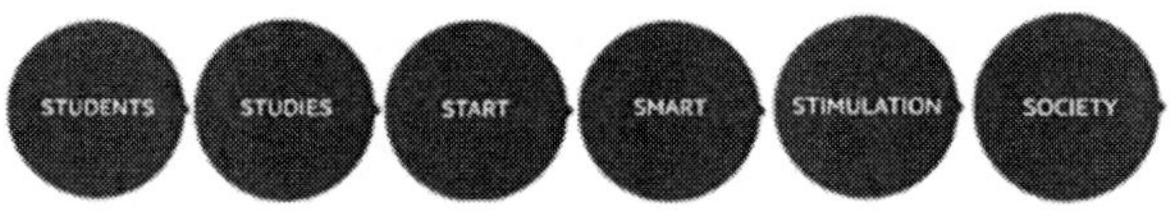

ITMO University was the first of all the 26 Russian "5-100" universities-participants to create a student strategic office of the program – the student subdivision "5-100S". Where "S" stands for:

- STUDENTS: The "5-100S" team is formed by students, first of all, master's and PhD students.
- STUDIES: Students learn to work on an international level in scientific, educational, innovative and management spheres solving problems in all the working areas.
- SMART: Non-classical university goes for non - classical solutions.

- START: School for young leaders. Recruitment of talented undergraduate and graduate foreign and home students onto ITMO University's educational programs.
- STIMULATION: Improving the university students and their interest to the university work.
- SOCIETY: project is aimed at the formation of a loyal community inside and outside the university among students and graduates of the university

Goals

- Informing youth audience about 5-100 Project and developing internal loyalty to the program.
- Realization student's initiatives in their science research works, education, innovation, PR and management activities.
- Incorporate student into the advertising activities of the university to attract students from all around the world and improve the integration process of foreign students and employees into the university environment and achieving effective cross-cultural communication.
- Involving students into the university's administration.
- Formation of highly qualified personnel for the university.

2. STRATEGIC INITIATIVES

The Roadmap of the "5-100A" office has the same structure and ideas that the main ITMO University Roadmap. Both of them include six strategic areas of development, which are "Science", "Education", "HR/Alumni", "PR", "Innovation", "Management - Students and University's future".

Figure 2. Student office 5-100s

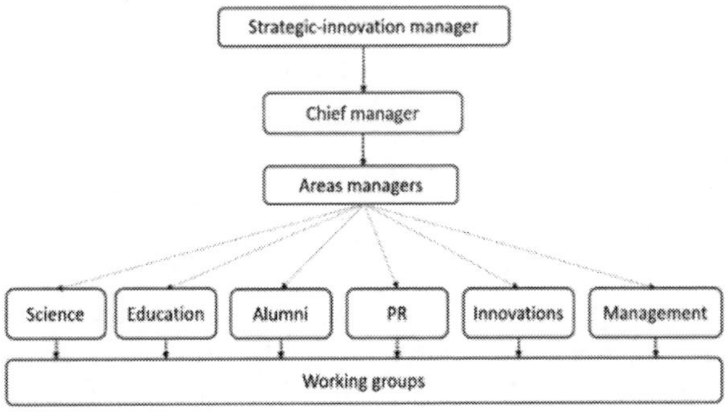

3. STUDENT PROJECT OFFICE "5-100S" STRUCTURE

The structure (fig. 2) of the Student Office "5-100S" is similar to the 5-100 Project structure, which includes the next components:

- Strategic Innovation Manager – it is the strategic manager of the entire 5-100 project, not only for the student office. It is the person responsible for the promotion of ITMO University at a global level, the formation of a strong international brand and the development of a loyal university community. The Strategic Innovation Manager knows where to go and what to do, that is the information given to the chief manager, who manages the student office to achieve all the Strategic Innovation Manager´s goals
- Chief Manager – should be an ITMO University student, responsible for the effective work of the "5-100S" Office and the integrations of the Student Office into the university development strategic plan. He/she is part of all the councils that take part in the 5-100 Project and is a representative of the 5-100 Project.
- Area's managers – are the coordinators of each working area of the 5-100 Office. The student Office "5-100S" has 6 working areas, those are:
  - Science (ITMO Youth Science)
  - Education (ITMO Education)
  - Alumni (ITMO Alumni)
  - PR (ITMO Student PR)
  - Innovations (ITMO Open Innovations)
  - Management (Students and University's future)
- Working group – 7 to 10 members of the Student Office, who count with their own projects and their own administrative, human and financial resources.

4. STUDENT OFFICE WORKING AREAS

The Roadmap of the Student office includes six strategic areas of development:

4.1 Science (ITMO Youth Science)

The main goal is to involve young people in an international scientific environment and increase the efficiency of the student's scientific activities in the university through:

- participation in international scientific societies (SPIE, OSA, IEEE and others);
- active involvement in the implementation of joint research projects with universities, leaders in international rankings;
- participate in competitions and grants (fundraising);
- realizing the project of open laboratories for foreign young scientists;
- material incentives for student publications

Main tasks:
1. Develop an environment for student international research activities and increasing ITMO University student's motivation to research activities.
2. Raising young people's publishing activity.
3. Promotion of ITMO University scientific achievements and in particular, the achievements of the university students and young scientists.

4.2 Education (ITMO Education)

The main goal is attracting foreign students to educational programs (short or double degree) into ITMO University from leading foreign universities by:

- realizing the project "ITMO Ambassadors";
- incorporation to international student associations, such as Erasmus Student Network (ESN);
- development of English-speaking environment;
- accomplishment of creative projects to attract foreign students to study;
- implementation of joint educational programs with leading foreign universities; (Institution research collaborations, double degree programs, summer programs, student exchange program)
- increase motivation among ITMO University students to study abroad;
- realizing the project of internships to ITMO university representative offices;
- implementation and promotion of popular subjects online course in English.

Main tasks:
1. Search talented young people and develop an enabling environment for the educational activities at the international level
2. Improving the quality of educational training, internationalization and promotion of the best educational projects and programs.

4.3 Alumni (ITMO University Alumni)

Main goal - building a loyal community of graduates including foreign students, involved in university life and contributing to its competitiveness development and growth by:

- participation of students in the implementation of alumni loyalty programs;
- organizing outdoor events for graduates;
- organization of alumni visits to ITMO University;
- attracting foreign graduates to work at ITMO University;
- promote ITMO University brand in the countries where ITMO University graduates live and work.

4.4 PR (ITMO Student PR)

Main goal - inform students about the Project 5-100, involving them in activities of "5-100S", increasing brand awareness and attractiveness of the university among young people through:

- Participation in big student international events, exhibitions, youth associations and student associations;
- Distribution of the "5-100S" Brand Book and its components among students for active use in all activities and working areas;
- Develop the students activities in social networks;
- popularization of science and innovations;
- increasing the number of publications in the Russian and foreign media;
- The development of a single ITMO University community and the promotion of the university corporate culture in the student's environment.

4.5 Innovations (ITMO Open Innovations)

Main goal - increase the capacity of the innovation field, creating a favorable environment and culture for the development of student entrepreneurship by:

- Organization of work and engaging students in public practice laboratories and other elements of the university innovation environment, for example, FabLab, etc.;
- organize meetings of students with well-known entrepreneurs and investors;
- participate in competitions of student grants and subsidies;
- Participating in international social entrepreneurship projects.

4.6 Management (Students and University's future)

The main goal is to involve students in university management processes, effective integration of the student office "5-100S" into the major 5-100 Project, through:

- formation of Student Office "5-100S";
- inclusion of students in solving strategic problems of the 5-100 Project through the implementation of the student's initiatives of six areas;
- increase the student participation in the university's councils.

Main task: administration of the "5-100S" Office and inclusion of students into university management process

5. PROJECTS REALIZATION: ITMO UNIVERSITY CASE

Since Student Project Office "5-100S" was recently created, all the projects and ideas that have been thought are just beginning to be implemented. Here are explained some of the ongoing projects of the Student Office "5-100S".

5.1 International Ambassador

ITMO Student Ambassador is a project in which ITMO University students will be able to prove themselves as the official representatives of the university and make a personal contribution to the promotion of the ITMO brand in Russia and abroad. The project aims to attract foreign students to study in Russia, involve students to the university′s corporate culture, as well as increasing the interest of Russian students to ITMO University.

ITMO Student Ambassador represents the university on external events and outside of ITMO University. ITMO's international ambassador is a person who knows all the necessary information about ITMO University, has good presentation skills and his main responsibilities are preparing presentations and negotiations for further partnerships between universities all around the world. That is why there is a special training program for ambassadors, where they learn how to present ITMO University to others. The selection of participants started in February 2016.

5.2 Academic board

ITMO University Academic Board since January 2016 includes 24 students. Twelve of these students are the presidents of their student's organizations, such as the "Foreign Student Council", "Campus Student Union", and "Students' Scientific Society" and so on. The other twelve student-members are chosen by elections, each faculty should chose its representative among all the faculty students, so only the best student that suits the place is selected from all the faculty′s students. All 24 student-members of the Academic Board are not simple students; they all have an experienced working background and know perfectly what problems the students have.

Over time students change, every new academic year welcomes new students with new ideas and new interests. University employees sometimes cannot adapt to these changes, that's one of the reasons why there are students in the Academic Board. Who knows their problems and their interests better, if not the students themselves?

Having students in ITMO University Academic Board giving their opinions, voting and proposing new projects will improve student's experience and their participation solving university issues as well as improving the working effectiveness of the Academic Board at all levels.

6. CONCLUSION

The goal of the Program for Enhancing the Competitiveness of ITMO University amongst the leading world's universities, research and educational centers (5-100) is to achieve a leading position among the world educational and research elite implementing these six initiatives:

- "Securing the World-class Level of Scientific Research and Development Projects in the Fields of IT and Photonics"
- "Global education: Personal Development and Professional Competitiveness"
- "Creation of a Corporate Culture and Environment that Positively Impacts on the Development of Highly Professional Personnel"
- "ITMO University Global Brand-building"
- "Growth of the Innovation Ecosystem Potential: Knowledge and Technology Transfer"
- "Transformation and Development of the University Management System",

and on this basis ITMO University aims to develop a highly skilled personnel in the interests of improving their competitive advantage.

Designating the long term plans for the project 5-100, ITMO University aims to create a highly professional management talent group among active motivated young students who will lead the university in the future, for whom it is important to create conditions for their development in the university′s framework.

Student Project Office "5-100S" aims to promote the university among the international community through the involvement of students in all of the university's key activities: science, education, innovation, PR, HR and management.

That is how ITMO University following shared governance concept, aims to train and draw students to a new process to achieve objectives and strategic goals for the university.

REFERENCES

- Shared Governance for Academic Units: Office of the Provost, October 2014. Retrieved form: https://provost.illinois.edu/communication/27/Communication_27_Shared_Governance_for_Academic_Units.pdf
- EXECUTIVE SUMMARY of the Action Plan of the Saint Petersburg National Research University of Information Technologies, Mechanics and Optics. – 2014. Retrieved form: http://5100.ifmo.ru/file/pages/9/roadmap_summary_itmo_eng_2014.pdf
- Complete Roadmap of ITMO University. -2014. Retrieved form: http://5100.ifmo.ru/file/pages/9/itmo_university_roadmap_presentation_2014.pdf
- Bochkova O., Zemnuhova L., Zhidkova N., Popova E., Shilo L., Chernish A. (2015). Shared governance in university. European University in Saint Petersburg.
- Rating Universities on Internationalization: QS Stars http://www.topuniversities.com/qs-stars/qs-stars/rating-universities-internationalization-qs-stars

Measuring Subjective Well-being among Youth in Tunisia: A Comparative Study across Gender and Regions

Molka Abassi
MSc in Economics Student, HEC Lausanne, Switzerland

1. INTRODUCTION

"Wellbeing cannot exist just in our head. Wellbeing is a combination of feeling good as well as actually having meaning, good relationships and accomplishment."
Martin Seligman[1]

Around 30% of the overall Tunisian population is aged between 15 and 29 years. This situation is characterized by many economists as a 'Youth Bulge'. Additionally, Tunisia's GDP has increased dramatically over the last decade. On the other hand, income gap has been widening and thus more people have fallen as victims of poverty, especially, in the hinterland regions of the country. Although the middle class has mushroomed in a generation, going from 70.5% in the 90's to around 80% in 2005, the gap between the rich and the poor has significantly increased. There is likewise a misconception that inequality in terms of development in Tunisia is not as high as in other Arab societies. The Gini coefficient[2] reveals similar results across Arab countries. Touhami (2014) reported that the Gini coefficient for Tunisia is about 39.8 against 35.3 in Algeria and 39.5 in Morocco, which are neighboring countries. Although the results do not appear to be very alarming, they are not far from results of countries considered to be very unequal such as Brazil with a coefficient of 51.9 or South Arica (63.1) than to countries that are believed to advocate equality like Sweden (0.26) according to the same source.

In addition to this regional inequality, the issue of youth unemployment is a long way from being solved. Young people aged between 15 and 30 years and who represent about one third of the total labor force, suffer from a high rate of unemployment of about 30%. This rate is even more pronounced for young graduates and females especially in the interior regions of the country Boughzala (2013). The reason for this very high rate of unemployment is mainly due to a mismatch between the supply and demand of labor: government failure to create jobs and attractive opportunities for youths on one hand and a deterioration of educational quality due to an increase in the number of students attending university on the other hand. High rates of poverty, regional disparities and unemployment coupled with high rates of corruption during Ben Ali's regime generated a sense of discrimination, exclusion and life dissatisfaction especially among the youth living in the western regions of the country. As Veenhoven (2000) noted, 'Freedom emerges from frustration'. In this particular context of regional disparities, youths from the interior part of the country started voicing their anger and indignation triggering the Arab spring in Tunisia in 2011 which then spread across the MENA region[3] and led to the fall of the dictatorial regime.

Tunisia is presently in a transitional phase of political, economic and democratic reforms. In October 2011, Tunisia witnessed its first democratic elections. These elections resulted in the formation of a constituent assembly, which was elected for the purpose of drafting a new constitution for the country that was passed on the 27th of January 2014. It has been described as the most progressive constitution in the Arab world (The Guardian 2014).

On another hand, many Tunisians have reported a deterioration of their financial situation in all income level groups. Thus, more pressure has been placed by Tunisians on the government particularly by the youths to improve public goods and services mainly educational institutions in hope of ensuring a smooth transition from adolescence to adulthood. Unlike in many Western countries, Tunisian youth remain in their parents' home until they marry. This is even more prevalent for females. Employment after schooling during this period is crucial especially for males to enable them to save money to cover marriage expenditures and housing costs which are the main costs endured by married couples especially in North African countries (Boughzala 2013).

Income and employment status have been designated as the principal determinants of overall wellbeing (Clark & Shields 2008). It is important at this stage of the introduction to define this notion of wellbeing. MacKerron (2012) quoting a work elaborated by

[1] Martin Seligman is an eminent American psychologist known for his 'learned helplessness' theory.
[2] Gini coefficient is a scale that ranges from 0 meaning perfect equality to a 100 meaning perfect inequality within a same country.

[3] MENA: Middle East and North Africa region.

Dolan et al. (2006), has distinguished five broad ramifications for the word wellbeing which are 'preference satisfaction, objective lists, flourishing, hedonic and evaluative'. Throughout this paper, I will use both terms 'Subjective Wellbeing' (SWB) and 'wellbeing' interchangeably to refer to only two main wellbeing concepts which are 'happiness' and 'life satisfaction'. Numerous economists have used the terms happiness and life satisfaction conversely like Veenhoven (2000) while Kahneman and Deaton (2010) separated the two concepts. Happiness or affective wellbeing has more to do with emotional wellbeing and the intensity of experiencing different emotions like anger, joy, stress and laughter that make an individual's life agreeable or disagreeable. Life satisfaction, on the other hand, translates thoughts that people have about their overall life. Income and Education are therefore more suspected to influence life satisfaction whereas loneliness and smoking look as if they affect happiness more.

Tunisians' wellbeing reportedly plummeted two years before the beginning of the uprising in 2008 (Clifton & Morales 2011). Consequently, Tunisia which used to rank on a highest ladder compared to other MENA countries in terms of wellbeing was found at the bottom of the same ladder in 2011. Wellbeing is not just a byproduct of life satisfaction and happiness determinants, but can also act as a moderator of one's behavior, emotion and cognition. It contributes to fostering social contracts and motivating people. While assessing GDP growth and other measures of economic performance is important, recent literature highlights the equal importance of tracking people's wellbeing and perceptions based on behavioral Economics and Psychology.

This paper seeks to add to the empirical evidence on the wellbeing literature. Studies on wellbeing in the Arab World and more specifically on Tunisia have been fairly uncommon. Thus, there is fertile ground for such an empirical investigation. This study will measure youth wellbeing across gender using fairly recent data issued after the 2011 upraising. For a long time, Tunisian women have been regarded as the most emancipated in the Arab Muslim world, (Tchaïcha & Arfaoui 2012). The Personal Status Code, which was promulgated in 1956 and considered to be the most liberal code in the region, granted Tunisian women numerous rights like forbidding polygamy and outlawing the husband unilateral repudiation (Tessler et al. 1978). Although women have actively contributed to the preliminary success of the so called 'Revolution for Dignity', changes that followed have brought new challenges for women, especially with a growing conservative philosophy spreading throughout the country (Tchaïcha & Arfaoui 2012). Consequently, it would be useful to measure the effect of these new challenges on young females' wellbeing. Finally, an attempt to discover the main determinants of youth wellbeing across different regions may help to suggest efficient policies for the diminishment of tension within the country.

The rest of this paper is structured as follows. Section 2 describes the Tunisia Household Survey on Youth in Urban Areas (THSYUA) dataset, gives an overview of the main differences between the regions and presents characteristics of the sample analysed in this study. Section 3 reviews the main results. The final section suggests policy implications based on the results, caveats of this study and thoughts on further research.

2. DATA & DESCRIPTIVE STATISTICS

2.1 Data

Tunisia's territory is shared between 66.2% urban areas against 33.8% rural, (INS 2012). The country accounts for 24 governorates aggregated into seven different administrative regions. In general, the North West, the Middle West and the South West are referred to as the interior part of the country whereas the Tunis District, The North East, the Middle East and the South East are referred to as the coastal regions.

The current paper uses a survey with information on inequalities between people in the whole Tunisian territory. It emanates from collaboration between the following three entities, the Tunisian National Statistical Office (INS), the General Commissariat for Regional Development (CGDR), and the World Bank (WB).

2.2 Regions Characteristics

7 million out of 10.6 million Tunisians (i.e. more than two-thirds of the population) reside in the country's cities and towns. As previously mentioned, the urbanization rate for the whole country is close to 62%. But this percentage varies a lot depending on areas. This is probably due to the natural advantages for market access and trade for coastal areas compared to the hinterlands. For instance, in the Tunis district almost 92% of the area is urbanized compared to 33.3% in the Middle West. Also, the overall population size differs, Tunis district accounts for 23.2% of the overall population against only 9.1% in the South East. Regarding socio economic characteristics, Poverty rates also differ significantly

between regions. They are three times higher in the poorest regions: 9.1% in Tunis District against 25.7% in the North West and 32.3% in the Middle West[4] in 2010.

Tunisia accounts for around 2.6 million households overall[5] but they are not evenly allocated between regions. As an example, Tunis district comprises 621.4 thousands households against only 214 thousands in the South East. The household composition remains approximately the same across regions. On average a Tunisian household is composed of four members, the parents and two children. With reference to durable goods, some striking findings are worth to be mentioned. Computers are only available in 25.7% of the overall Tunis district households against only 6.8% in the North West and a very modest 3% in the Middle West. Regarding Internet coverage, there are also some disparities. There is about 28% Internet coverage in Tunis district against only 7% in the Middle West and 9.9% in the North West.

On the individual level, there are also many observations to draw. Around 6.2% of the Total Tunisian GDP is allocated to education, which places the country on the 38th position worldwide[6]. Moreover, education is free in Tunisia and considered as one of the best in the Middle East and North Africa. Despite these facts, illiteracy rates remain high especially among women in certain regions. As a matter of fact, the North West region registers a rate of illiteracy of 40.5% for females and the Middle West a rate of 41.7%. For males, rates are relatively lower.

To conclude this part, it is important to note that disparities between regions are undeniable. Regions differ on the infrastructure, technology, human capital, demographics and natural characteristics levels. Females living in the poorest regions are the least fortuitous possibly because of social constraints imposed on them by the society. Youths living in the hinterlands of the country are also disadvantaged on all levels. Lack of opportunities is likely to affect their overall wellbeing and their future aspirations. It has also even produced a fertile ground for the rise of extremism in these regions especially after the revolution, (Touhami 2014). The first and most urgent priority for the development to be sustainable is not to directly invest in projects in the disadvantageous regions but to nurture Tunisia's human capital.

[4] *National Institute of Statistics, National Expenditure Survey, Households consumption and living standards, 2010.*

[5] *2692 thousands households ,according to the National Institute of statistics, INS, 2011.*

[6] *CIA World Factbook*

2.3 Key Variables

2.3.1 dependent variables

This study is looking at two dependent variables LS, which stands for Life Satisfaction and Happiness. Both variables are ordinal in nature and are based on how the respondent answers the following two questions: 'Generally do you consider yourself unsatisfied, unsatisfied to some extent, satisfied to some extent or very satisfied?' and 'Generally do you consider yourself unhappy, unhappy to some extent, happy to some extent or very happy?' The mutually exclusive responses are coded from 0 to 3, 0 being the lowest bound and 3 being the highest one. The sample is restricted to youths aged between 15 and 29 years.

2.3.2 explanatory variables

Many regressions' specifications have been carried out using information from the data and the most relevant variables have been retained for the main specifications' models.

Covariates for the two models can be classified into 3 main categories: individual, household and region specific variables. The individual variables, which add up to 17, contain personal characteristics variables like age, marital status, personality traits, trust variables, education and health status variables. The individual variables also include weekly activities like religion, exercising, and spending time with friends or on the Internet and socio economic variables like employment status. In regards to household specific variables, both main specifications include household size; wealth indicators and household log income. With reference to the last section, regions dummies are added to the main models to capture their fixed effects. The covariates variables add up to 30 in total.

3. RESULTS AND DISCUSSION

3.1 Results

Table 1 above displays the results from estimated Life Satisfaction (LS) and Happiness Models. Models have been chosen after a battery of diagnosis to check for their adequacy. Column 1 presents results for the pooled LS model; column 2 considers only male individuals and column 3 considers only females (same for the happiness model in the next three columns of the same table). This distinction has been made to disentangle gender differences.

Table 1. Maximum Likelihood Estimates for Ordered Probit Models

Standard Ordered Probit Models						
Variables List	**LS Model**			**Happiness Model**		
	Pooled	**Males**	**Females**	**Pooled**	**Males**	**Females**
	1	**2**	**3**	**4**	**5**	**6**
Individual specific variables						
age	-0.06	-0.1	-0.05	-0.12	-0.28**	0.01
age2	0.00	0.00	0.00	0.00	0.01**	0.00
married	0.45***	-0.01	0.60***	0.51***	-0.01	0.63***
male	-0.05	0	0	-0.14*	0	0
confident	0.51***	0.51***	0.47***	0.37***	0.47***	0.21
independent	0.32***	0.37***	0.32**	0.33***	0.29***	0.40***
sociable	0.17*	0.16	0.17	0.24**	0.33**	0.17
timereligion	0.1	0.23**	-0.01	0.12*	0.21**	0.02
timeinternet	0.06	0.1	0	0.06	0	0.15
timephysiac	-0.04	0	-0.14	-0.03	0	-0.11
timefriends	-0.01	-0.08	0.05	-0.01	-0.14	0.1
pastinc	0.18**	0.20*	0.18	0.18**	0.09	0.29***
truststate	0.24***	0.18*	0.29***	0.19**	0.17	0.22**
trustschool	0.06	-0.1	0.26**	0.17**	0.09	0.30**
trustimams	0.1	0.08	0.14	-0.01	0.04	-0.04
trusgovoff	0.24***	0.39***	0.12	0.13*	0.19*	0.07
satishealth	0.87***	0.82***	0.99***	0.71***	0.73***	0.72***
inschool	0.27***	0.51***	0.08	0.1	0.38***	-0.11
unemployed	-0.21***	-0.37***	-0.07	-0.12	-0.41***	0.19
Household specific variables						
inhhinc	0.08*	-0.02	0.19**	0.11**	0.02	0.17**
sizefamb	0.15**	0.09	0.25***	0.11*	0.06	0.21**
dw_high	0.27***	0.34***	0.23**	0.17**	0.11	0.28***
car	-0.06	-0.05	-0.09	0.23***	0.32***	0.13
computer	0.24***	0.17	0.36***	0.08	0.23**	-0.04
Region specific variables						
regNorth_East	0.02	0	0.07	-0.09	0.01	-0.2
regNorth_West	-0.22**	-0.24	-0.18	-0.29***	-0.59***	-0.05
regMiddle_East	-0.18**	-0.33***	-0.04	-0.08	-0.12	-0.1
regMiddle_West	-0.23**	-0.19	-0.32**	-0.26**	-0.38**	-0.21
regSouth_East	0.17	-0.11	0.31*	-0.01	-0.30*	0.07
regSouth_West	0.20*	0.12	0.28*	-0.09	-0.18	-0.05
cut1	0.04	-0.83	0.85	-1.01	-3.12**	0.64
cut2	0.54	-0.32	1.35	-0.34	-2.35*	1.24
cut3	2.16**	1.24	3.08**	1.92**	-0.11	3.62***
N	1686	795	891	1686	795	891
pseudo R-sq	0.112	0.119	0.126	0.1	0.123	0.1
"* p <0.10 **p <0.05 ***p <0.01"						

3.1.1 discussion

Generally, the results provided in Table 1 are compatible with findings in section 2 with some notable twists reflecting characteristics peculiar to an Arab Muslim society. Being a male has a negative impact on overall happiness. Age does not appear to have a significant influence on the wellbeing variables, potentially because of the relatively small age gap in the sample. Regarding gender differences in happiness, women appear to enjoy greater happiness levels despite the fact of being disadvantaged compared to young males. This could potentially be explained by the fact that young females in general are more optimistic and have a better life balance than their counterparts. This result is consistent with the literature as Plagnol and Easterlin (2008) argued that women are likely to report high happiness levels than male in their early adult life. Young females are more likely to satisfy their family aspiration and their need for material goods than their counterparts according to the same source. Additionally, being married appears to only affect young females' life satisfaction and happiness positively. Although early marriage has declined in most Arab countries and especially in Tunisia, it is still considered a turning point in a bride's life and a joyful celebration for the whole family (Rashad et al. 2005). The effect of marriage is maybe not as pronounced for males due to the small percentage of married males in the sample. Regarding personality traits, confidence and independence emerge to positively affect both genders' life satisfaction and happiness, which is in accordance with literature.

99% of Tunisia's population is Sunni Muslim. Although, women on average reported spending more time practicing religious activities during the week prior to the interview date religiosity appears to only affect males' happiness and LS positively. This finding contradicts Abdel-Khalek and Naceur (2007) research on Algeria when he reported that religiosity contributed to increase wellbeing for women more than it did for men. A potential explanation of this result is that religion for males facilitates social integration (Barakat 1993). In Arab Muslim communities, 'society is the soul of religion' according to the same source. Mosques in addition to being pious places, are meeting points, especially for males, from all society classes.

Regarding trust variables, trust in state institutions happens to be positive and significant for both genders unlike Veenhoven (2000) findings. In Tunisia, political as well as economic freedom mater for both genders. For women, trust in state institutions is significant at the 1% level. Women have contributed significantly to the 'preliminary' success of the revolution because they effectively believed that state institutions could be improved. Even during the aftermath of the revolution, women took to the streets along with men to defend their rights and the rights of Tunisia's future generations, especially when they feared that Islamist led governments would roll back their rights. Additionally, trust in schools is positive and significant at a level of 5% but only for young females. This finding could be explained by the importance placed by girls on education. As outlined in previous sections, female enrolment rates in all levels have risen, translating a real hope for emancipation. Moghadam (2003) affirms that young females associate a better education with a higher social status, a higher income and thus more economic freedom vis-à-vis of husbands. Trust in government officials in contrast, only affects males' wellbeing positively. This could be attributed to the monopolization of political and administrative positions by the male population. For instance, even though both genders should be equally represented in the constituent assembly, only 67 females have received seats out of 217 in total[7]. Comparable observations can be drawn for the proportion of females' participation in the current government composition.

Living in big sized families composed of six members or more, emerges to only affect women's wellbeing positively. Analogous findings are highlighted in Oliva and Arranz (2005). According to Joseph (1994), families are a core unit in Arab societies. Enjoying solid inter-sibling exchanges and social support from older brothers and sisters could diminish the impact of the patriarchal environment imposed by parents on their daughters.

On another hand, economic characteristics as expected, affect both genders. However, remarkable results emerged to reflect characteristics idiosyncratic to an Arab society. The negative effect of unemployment pushes young individuals back down the real line towards lower levels of LS categories. Males place a high value on work. Having a decent job promotes social integration and improves social status, which are two important determinants for males' satisfaction. Another hypothetical explanation for this result is that males, in most cases, aspire to being breadwinners, a position where the entire burden of the family is placed on them. Rama (1998) demonstrated that the unemployment rate of households' heads was as low as 1.4% in 1990 in Tunisia[8]. This percentage is minor

[7] El Marsad website.

[8] A more recent figure for households' heads rate of unemployment is unfortunately not available to display.

compared to first-time job seekers like graduate students. Unemployment among graduates can be considered as 'voluntarily' unemployment according to the author because young individuals take advantage of their family support to wait for the right job opening, while rejecting existing work opportunities that are not very attractive. In contrast, households' heads feel that they have the responsibility to feed and take care of their families and accept even the least rewarding jobs.

While unemployment emerges to have a strong negative effect on young males, the effect of unemployment on young females does not appear to be a strong determinant of their wellbeing. Contrary to this finding however, income appears to only affect young women wellbeing positively. This can be explained by the fact that a majority of Tunisian women manage the household budget and expenditures. Having an additional income would therefore increase their overall wellbeing. Also an increase in relative income (i.e. current household income relative to past incomes) affects positively the young individual overall wellbeing. The effect is even more pronounced for the females' happiness model. Additionally, wealth related variables like living in a high quality habitat appears to also affect wellbeing positively. This result is again in line with a previous work elaborated by Cummins (2000). The authors provided evidence that money actually buys happiness by comparing rich and poor individuals.
Regarding the regions' dummies for the LS model, the results suggest that on average and ceteris paribus, the North West, the Middle East and the Middle West regions of the country place an average youth further down the real line compared to the base group, which is Tunis district in the case of these models. Although the results for the North and the Middle West were expected, the negative effect of living in the Middle East on LS was not anticipated as this region is considered among the most developed regions of the country. This result could potentially be explained by the increase of youth migration from the western parts in search for work in the Middle East region, which provides better employment opportunities. Youths living in the Middle East region feel maybe that these migrants compete with them to get jobs.

4. CONCLUSION

The revolution, triggered by Tunisian youths in 2011, succeeded in bringing down dictatorship peacefully. This demonstrates the power the youths possess to positively benefit the country. It has therefore become imperative for policy makers to recognize this huge potential and empower the youths by investing in human capital

The overall wellbeing of a country can no longer be isolated from that of its younger population. Using a rich dataset from Urban Tunisia issued after the 2011 Revolution, this paper studied the determinants of wellbeing among youth across gender and regions with a particular emphasis on social and economic variables trust in different institutions, religiosity and personality traits. In accordance with the literature on SWB in the Economics field, this study has used an Ordered Probit model to measure both youth life satisfaction and happiness. While most of the studies in the literature do not attempt to disentangle differences between the two concepts and instead using them interchangeably, this study was able to measure both their determinants separately. Although different effects were expected, the covariates have shown similar effects for both the happiness and the life satisfaction models. This result goes along with Blanchflower and Oswald (2004) when measuring these two concepts for the UK and the USA.

While the results are for the most part in accordance with existing literature, there are some twists reflecting specificities distinctive to an Arab Muslim society. On the individual level, Tunisian young females reported on average higher levels of happiness than males despite the patriarchal composition of the society. Religiosity happens to affect positively males while marriage on the other hand was reported as a potent determinant of females' wellbeing. The effect of marriage for males is maybe not as pronounced due to the low percentage of married males in the sample.

Additionally enjoying good health appears to be a very important determinant for overall wellbeing for both genders. Personality traits as well are found to positively affect both genders with sociability being only significant for males' happiness. This again translates the patriarchal nature of the Tunisian society. Tunisian educators, reformers and psychologists must take into account the multidimensional nature of the SWB predictors and be aware of the implications of wellbeing for positive development and good mental health.

Household relative and absolute income have emerged as a strong predictor of overall wellbeing especially for women. As income increases, the level of wellbeing improves systematically. This implicitly stresses the importance of employment as a sustainable source of revenue. It also emphasizes on the government and the private sector roles to create and secure jobs for the thousands of unemployed Tunisian graduates.

Trust in state institutions as well, is a strong predictor of wellbeing translating into a real desire of youths for institutional improvements. Consequently, today's government should foster youth civic engagement and involve the younger population in planning development reforms by creating suitable platforms for youths that enable them to become proactive players in their respective regions. The government should also be cognizant of the importance of the decision taken as they have direct impact on youth trust. Any decision reflected poorly can cause a loss in youth confidence and can lead to the emergence of new conflicts.

Additionally, it was clearly noticed throughout the paper that the North and the Middle West regions have reported lower levels of wellbeing due to low levels of development in these specific regions compared to their counterparts. However, contrary to what was expected, the southern part of the country has reported positive levels of life satisfaction among females. Further investigation of potential explanations of this finding could be part of an agenda for future research. Although the main specifications have given relatively reasonable results, there are some caveats that need to be mentioned. First, data only included urban areas in Tunisia. Excluding rural areas restricted the sample to only youths living in the country's towns and cities. Although rural areas only account for one third of the country, their inclusion would have explained even more differences in wellbeing across the country and would have allowed for a total representation of the youth population. In addition, although the THSYUA dataset has covered many important variables, it didn't include an extensive array of objective measures of education, health and consumption of cultural products, which are strong determinants of youth wellbeing. There is therefore a potential omitted variable bias affecting the models' specifications. However this does not in any way undermine the importance of the study findings and it provides a good foundation for future research to build on.

REFERENCES

- Abdel-Khalek, A. (2004). Research Note: Happiness among Kuwaiti College Students. *Journal of Happiness Studies, 5,* 93-97.
- Aitchison, J. and Silvey, S. D. (1957). The generalization of probit analysis to the case of multiple responses. *Biometrika,* 131-140.
- Al-Qudsi, S. S. (1998). Labour participation of Arab women: estimates of the fertility to labour supply link. *Applied Economics, 30(7),* 931-941.
- Arab Social Media Report (2011). Facebook Usage: Factors and Analysis. *Dubai School of Government, 1(1),* 1-20
- Ayadi, L., Benjamin, N., Bensassi, S., and Raballand, G. (2013). Estimating Informal Trade across Tunisia's Land Borders. The World Bank Middle East and North Africa Region Public Sector and Governance Unit, Policy Research Working Paper 6731.
- Ayyash-Abdo, H. and Alamuddin, R. (2007). Predictors of subjective well-being among college youth in Lebanon. *The Journal of Social Psychology 147(3),* 265-284.
- Blanchflower, D. G. and Oswald, A. J. (2004). Well-being over time in Britain and the USA. *Journal of public economics, 88(7),* 1359-1386.

- Blanchflower, D. G. and Oswald, A. J. (2004). Well-being over time in Britain and the USA. *Journal of public economics, 88(7),* 1359-1386.

- Boughzala, M. (2013). Youth employment and economic transition in Tunisia. Global Economy and Development, Working Paper 57.
- Cummins, R. A. (2000). Personal income and subjective well-being: A review. *Journal of Happiness Studies,1(2),* 133-158.
- Frey, B. S. and Stutzer, A. (2002). What can economists learn from happiness research? *Journal of Economic literature,* 402-435.
- Frey, B. S. and Stutzer, A. (2002a). Happiness and Economics: How the Economy and Institutions Affect Human Well-Being. Princeton University Press.
- Gilman, R. and Huebner, E. S. (2006). Characteristics of Adolescents Who Report Very High Life Satisfaction. *Journal of Youth and Adolescence, 35,* 293-301.
- Graham, C. (2009). Happiness around the world: The paradox of happy peasants and miserable millionaires, Oxford University Press.
- Graham, C. and Chattopadhyay, S. (2012). Gender and Well-Being around the World: Some Insights from the Economics of Happiness. Working Paper No. 2012-010.
- Hattab, M. (2013). Les migrations intérieures et le risque social en Tunisie L'economiste Maghrebin.
- Headey, B., Veenhoven, R., and Wearing, A. (1991). Top-down versus bottom-up theories of subjective well-being. *Social Indicators Research, 24(1),* 81-100.
- Homel, R. and Burns, A. (1989). Environmental quality and the well-being of children. *Social Indicators Research, 21,* 133-158.

- Hudson, J. (2006). Institutional Trust and Subjective Well-Being across the EU. *Kyklos, 59(1),* 43-62.
- INS (2012). The National Census on Individuals and Employability 2011.
- Joseph, S. (1993). Connectivity and Patriarchy among Urban Working-Class Arab Families in Lebanon. *Ethos, 21,* 452-484.
- Litchfield, J., Reilly, B., and Veneziani, M. (2012). An analysis of life satisfaction in Albania: An heteroscedastic ordered probit model approach. *Journal of Economic Behavior & Organization, 81(3),* 731-741.
- Machin, S. J. and Stewart, M. B. (1990). Unions and the financial performance of British private sector establishments. *Journal of Applied Econometrics, 5,* 327–350.
- Mackerron, G. (2012). Happiness Economics from 35 000 feet. *Journal of Economic Surveys, 26,* 705-735.
- Ministry of Social Affairs (2014). Guide of Social statistics 2012, Bureau of Studies and Planning.
- North, D. C. (1989). Institutions and economic growth: an historical introduction. *World development, 17(9),* 1319-1332.
- Oliva, A. and Arranz, E. (2005). Sibling relationships during adolescence. *European Journal of Developmental Psychology, 2(3),* 253-270.
- Rama, M. (1998). How bad is unemployment in Tunisia? Assessing labor market efficiency in a developing country. *The world bank research observer, 13(1),* 59-77.
- Rashad, H., Osman, M., and Roudi-Fahimi, F. (2005). Marriage in the Arab world. Population reference bureau (PRB).
- Scales, P. C., Benson, P. L., Roehlkepartain, E. C., Hintz, N. R., Sullivan, T. K., and Mannes, M. (2001). The role of neighborhood and community in building developmental assets for children and youth: A national study of social norms among American adults. *Journal of Community Psychology, 29,* 703-727.
- Shallal, M. (2013). Job satisfaction among women in the United Arab Emirates. *Journal of International Women's Studies, 12(3),* 114-134.
- Suhail, K. and Chaudhry, H. R. (2004). Predictors of subjective well-being in an Eastern Muslim culture. *Journal of Social and Clinical Psychology, 23,* 359–376.
- Tessler, M. A., Rogers, J., and Schneider, D. (1978). Womens emancipation in Tunisia, 141-58.
- The Guardian (2014). Tunisia signs new constitution [Online].
- Touhami, H. (2014). Les deux Tunisie sous l'angle des inégalités de revenu Leaders.com.
- Veenhoven, R. (2000). Freedom and happiness: A comparative study in forty-four nations in the early 1990s. Culture and subjective well-being, 257-288.
- Winkelmann, L. and Winkelmann, R. (1998). Why are the unemployed so unhappy? Evidence from panel data. *Economica, 65(257),* 1-15.
- Zaiem, M. H. (2011). Les inegalites regionales et sociales dans l'enseignement superieur.

The Role of Mobility in Overcoming Labour Market Obstacles after the Financial Crisis

Elena Alcalde Peñalver, Ph.D.
Lecturer, University of Burgos, Spain

1. INTRODUCTION

We live in a globalized world in which importance is increasingly being given to the ability to communicate in two or more languages. This opens new doors to different opportunities for personal mobility, employment, education, and access to information. Accordingly, the Spanish government's Comprehensive Program for Learning Foreign Languages argues that, in order to achieve its planned objectives, it is necessary to raise awareness of the importance of learning foreign languages (Spanish Ministry of Education 2010).

We consider it appropriate to begin our paper with a reference to the theory of human capital (2003/C 295/05):

'Human capital is a strategic resource for the overall development of Europe as well as for the development of the education and training policies of the Member States. Therefore, they should aim to enhance the personality of each individual throughout their life and to achieve greater citizen participation in terms of social cohesion and economic development.'

'(...) The objectives defined in the policies of education and training take into consideration as one of their main purposes the harmonious development of young people to become self-confident, responsible, and cultivated citizens. They are increasingly complemented with economic and labor policies in order to combine social cohesion and competitiveness.'

As we can see, the theory of human capital is related to the principle of continuous education that prevails in the 21st century. Today, the idea of maintaining just one professional career throughout life is no longer taken for granted. Still, in some countries like Spain, this situation prevails, because working conditions in the public sector are far better than in the private sector at a quantitative as well as qualitative level. Thus, many young people who do not want to live and work in another country opt to take the qualifying exam for civil service employment, where working conditions will usually be better than at a private firm.

To improve the human resource potential of a country, investment in education and training is considered key by the European Union, as stated in the Council of Europe's Education and Training 2010 program:

'Human resources are the main asset of the European Union. They are fundamental for the creation and transmission of knowledge and a determining factor of the innovation potential of each society. Investment in education and training is a key factor of competitiveness, sustainable growth and employment of the Union.'

Explained in this manner, the training of human capital is related to the fulfilment of economic objectives, such as reducing unemployment or social inequality. As part of this training, as we will explain later, learning foreign languages is key.

2. THE CONCEPT OF COMPETENCE

The concept of competence in education is becoming increasingly important in the current analytic discourse of employability. This association of the concepts of competence and employability establishes a link between education and the labor market (Huertas 2013, p. 19). Thus, in this new educational paradigm within the European Higher Education Area (EHEA), Calvo (2005, p. 14, 2009, p. 150) sees the connection between professional profiles, learning objectives, and competencies as a cyclical conception that relates the various activities present or necessary in society and training.

The Framework Document on Spanish University System Integration in the EHEA (MECD 2003, p. 7) states :

'The training objectives of the official teaching programs at an undergraduate level will generally be undertaken with a professional approach, that is to say, they should provide university level training in which core generic skills, transversal skills related to the comprehensive training of people, and more specific skills enabling professional guidance are harmoniously integrated to enable graduates adapt to the labor market.'

'Competencies' are the basic elements on which professional qualification profiles have been analysed. The contents of the programs must be aimed at developing basic competencies at the undergraduate level. Subsequently, postgraduate training allows for more advanced specialization and mastery of skills required by the academic and professional profiles of these advanced degrees (Valencia 2006, 9).

Regarding the debate on the approach that education should follow, Gros (1995, p. 187) states:

'We should not be interested only in training individuals to enter the labor market. Instead, we should promote the full development of the individual in society. Higher education aims to provide comprehensive training to individuals so that they have sufficient resources for their permanent development throughout life.'

It is within this framework that we consider it necessary to implement changes so that students no longer consider languages something that they "must" learn in order to enter the labor market— that is to say, something imposed by the system—but rather, as a communication tool. In fact, if we consider the competencies that appear in the teaching guide of an English course currently taught at the University of Burgos in Spain, we observe that the curriculum takes into account the full development of the student as well as the acquisition of knowledge that enables mastery of the language needed for the student's profession. These competencies are:

Competencies English language 2, Education Degree, and English Mention:

- Students will be able apply their knowledge of English to their profession and will possess the competencies typically demonstrated when defending arguments and solving problems within their field of study.
- Capacity for analysis and synthesis.
- Ability to manage information.
- Teamwork.
- Critical Thinking.
- Autonomous Learning.
- Creativity.
- English communication skills involved in the exercise of the profession.
- Development of the ability to express themselves orally and in writing in a foreign language (B2 is the minimum level required).

We observe that this list of skills applicable to a particular subject is an example of the combination of knowledge and necessary skills for the job market as well as skills aimed at training the student at a more general level. The challenge for lecturers is how to combine the methodology of the class so that English moves from being a subject imposed on students in the education system to considering it a communication tool that is naturally acquired during their training.

3. CHANGES IN THE EDUCATIONAL SYSTEM IN SPAIN

In relation to the above-mentioned aspects, some changes have been implemented in the curriculum of Spanish schools. Spain remains among the countries with one of the lowest levels of foreign language proficiency. According to data from the 2015 English Proficiency Index (EPI), the countries of northern Europe had the highest English proficiency levels, whereas Spain was ranked number 23 within the bloc of countries with 'moderate proficiency.' Among European nations, only Italy and France lagged behind. Nevertheless, the 'theory' stated in the law of the Spanish education system from 2013 does give importance to foreign language proficiency as summarized below:

Proficiency in a second or even a third foreign language has become a priority in education as a result of globalization. However, it remains one of our education system's principal shortcomings. The European Union sets the promotion of multilingualism as an essential objective for the construction of the European project. The law strongly supports multilingualism, redoubling efforts to ensure that students are fluent in at least a primary foreign language, whose level of listening and reading comprehension and oral and written expression is crucial to promoting employability and career ambitions. Therefore, we are strongly committed to the incorporation of a second foreign language in the curriculum.

Taking this situation into account, it should be noted that for several years in Spain the emphasis has been on the need to improve language proficiency through what has been called 'bilingual centers' in public schools. However, these initiatives have been largely unsuccessful due to teachers' poor training. As Marias (2015) states, *'these teachers have a poor knowledge of the language, they usually have a heavy accent, or they ignore the pronunciation of different words, grammar, and syntax and tend to merely imitate Spanish language structure.'* It is very important that teachers be up to date both in education and in their professional fields in order to train their students on the basis of a set of previously defined competencies and learning outcomes (González & Wagenaar 2003, p. 50). Therefore, much more than just a simple language proficiency certificate should be required to teach foreign languages in public school. If we do not ensure the successful preparation and training of foreign language instructors, students will not be able to achieve the communicative competencies indicated by law.

For years, one of the European Union's main prioritieshas been for university policies to encourage teacher training through exchange programs such as Erasmus+, SUD-EU and other such programs at national and regional levels. These programs provide funding to conduct research or complete language residencies at other universities or schools within Europe or other continents. We will discuss these initiatives in more detail later on.

4. TRAINING AT THE UNIVERSITY LEVEL. WHY IS IT NOT ENOUGH?

Given the current economic environment, professionalization is one of the key features of university level training, since society shows great interest in the relationship between university education and integration into the labor market. We often hear professors and employers state that there is a major disconnect between these two groups, resulting in a large gap between what is required from workers and the preparation they receive before entering the labor market. The cause of this has traditionally been rooted in a lack of communication between universities and companies. Therefore, the need to promote greater communication between governments, higher education institutions, and employers constitutes a fundamental basis for conducting reforms (*London Communiqué* 2007, p. 5).

This gap between university education and labor market needs has made many university students increasingly opt for supplementary coursework to help them acquire needed skills, causing a progressive commodification of education and increased costs for students (Brunet & Belzunegui 2003). In order to illustrate this, and from my personal experience, although students may have taken English classes in college, they still choose to learn English in private schools because they do not consider themselves sufficiently prepared with the skills learned in higher education.

One could envision the reasons for this situation. On the one hand, the fact that university education in Spain has a fairly low cost compared to other countries such as the UK or the US could be influential, since in some cases students seem to take more seriously the courses for which they pay the most. On the other hand, it could be because more importance is being given to the practical usefulness of what is learned to integrate it into the labour market, which is in line with the statements of Yorke and Knight (2006, p. 4), who write that *'success will depend upon the extent to which students see a 'pay-off' for the effort they put in.'* This would relate to the quality of university level language teaching, which is perhaps not entirely focused on improving necessary language skills to be able to successfully confront different communicative situations.

5. HOW SHOULD WE TEACH LANGUAGES?

Given the new requirements imposed by the EHEA for higher education, we believe that the following objectives are essential to learning modern languages:

- From a communicative point of view, the student should be able to understand and produce speech acts that are appropriate to the communicative intention and the situation in which they are produced. When we talk about communicative competence the most important aspects are the adequacy and efficiency in the speech act (Berenguer 1996).

- Create working situations that resemble as faithfully as possible real communication situations for the student to feel that what they are working on can be applied outside the classroom.

- Propose a specific sequence for the teaching of modern languages with different themes that progress according to the level of specialization and which in turn incorporate the elements learned in previous units. This will help students assimilate the contents through repetition and the incorporation of different communication situations.

- Emphasize cooperative work as a learning philosophy that is based on the joint efforts of students who build their own learning experience in a social and real context (Tsokaktsidu 2005). Numerous studies have already demonstrated the effectiveness of cooperative learning in terms of improving student achievement, motivation for learning, and a sense of responsibility (Diaz Aguado 2003).

- Encourage self-learning on the basis of teamwork. This way, students will become aware of their own limitations, decide their own pace of work and will be responsible for their own learning (Berenguer 1996). This will allow students to continue practicing with their workgroup outside the classroom to work on those aspects that may have not been completely assimilated.

- To develop communication skills, increase participation and coordination between groups and thus facilitate motivation, self-esteem, and support among students.

6. THE ROLE OF THE EUROPEAN UNION

Mobility of staff, students, and graduates is one of the core elements of the Bologna Process, creating opportunities for personal growth, developing international cooperation between individuals and institutions, enhancing the quality of higher education and research, and giving substance to the European dimension (*London Communiqué* 2007). Therefore, the European Union has served a very important role in enhancing mobility programs.

First of all, experiences that all Europeans share include the continent's history and a unified interest in sharing power, improving production, and sharing common experiences based on a solid foundation of knowledge. This has been one of the main causes of European society's push to reorient educational processes and make their system more competitive (Pérez 2015).

The European Union, through its various agencies (European Parliament, Council of the European Union and European Commission), has created various commissions to develop strategies and mechanisms for academia, such as the Committee on Culture and Education of the European Parliament, which is responsiblefor coordinating the EU's educational policies such as the Erasmus program. The Council of the European Union, through its Education, Youth, Culture and Sports Council, promotes student and faculty mobility. It is also important to note the work of the European Commission, through the agency of Education, which focuses on the development of education systems. Thus, the European Commission makes annual investments in academic exchange programs through the Erasmus+ program.

Despite these efforts, a decade after the Bologna process began, the 2009 Leuven and Louvain-la-Neuve Communiqué stated that, although the results were satisfactory, not all objectives had developed to the same level. The report projected that by 2020 at least 20% of the European population with a degree would have had either a training or educational experience abroad. In the latest report called Implementation of the Bologna Process (2015), more updated information was presented, including an increase in the percentage of young people with experience in several countries. Although there is still insufficient data to demonstrate compliance with the target set for 2020, the current data mainly shows that most countries in the Bologna Process have experienced a rate of incoming and outgoing participants in mobility programs in the territory of the EHEA at or approximately 10%, with more than half of the countries presenting rates below 5%. The main obstacles to mobility remain bureaucratic obstacles and lack of funding and recognition of study periods abroad. Both students and staff agree that quality is more important than quantity and that more should be done to invest in information systems and control of learning experiences.

The Erasmus Impact Study commissioned by the European Commission in 2014, underscored the tremendous benefits of the Erasmus program. One of the great effects of Erasmus mobility has been the impact this has had on the dissemination and learning of the most important European languages. The lingua franca remains English, with 67% of students reporting that classes at their host universities were taught in this language. Additionally, this study highlights that since the beginning of the Erasmus program, more than 3 million students and more than 300,000 teachers and other university staff have benefited from this mobility program.

If, apart from these data, we consider other studies such as EURODATA Student Mobility in European Higher Education, ERASMUS statistics, the Flash Eurobarometer reports, and the study "Mapping Mobility in European Higher Education," funded by the European Union, it is clear that the experience of academic mobility of European students not only enriches their professional and academic life, but also improves the acquisition of a second language as well as their intercultural skills. In addition, after the completion of an Erasmus stay, different studies have demonstrated that academic mobility experiences increased the employment chances of 52% of participants.

7. CONCLUSIONS

The European Union cannot impose a common education policy for all member states, however, the various forums organized for the exchange of ideas between institutions in different countries and other similar initiatives have contributed to the establishment of principal courses of action (Madrid 2007). In this regard, we believe that exchange programs of the European Union should be key initiatives to encourage foreign language learning and networking for both instructors and students. It is clear that the learning of foreign languages is emphasized in European laws, initiatives, and treaties both at national and European levels, but there is still a long way to go.

In conclusion, it is common to hear that mobility enhances cultural and communication skills and that it is highly valued by employers. The financial crisis and lack of professional opportunities for many people around the world have made many of us begin thinking of mobility as something more than just traveling to another country to learn its culture and language. We consider that mobility should not just be an option to be considered, but rather, expected of any student, educator, researcher or professional. Thus, we consider that increasing mobility will result in the development of professionals with global perspectives that will understand each other better and see opportunities where others see boundaries.

REFERENCES

- Berenguer, L. (1996). Didáctica de segundas lenguas en los estudios de traducción. Amparo Hurtado Albir (Ed.), La enseñanza de la traducción, Castelló de la Plana: Publicacions de la Universitat Jaume I.
- Brunet, I. y Belzunegui, A. (2003). Flexibilidad y formación: una crítica al discurso de las competencias. Barcelona: Icaria.
- Calvo Encinas, E. (2005). *El paradigma de las competencias en enseñanza y aprendizaje y su aplicación a los estudios de traducción: el ejemplo del Reino Unido.* II Congreso Internacional de la Asociación Ibérica de Estudios de Traducción e Interpretación (AIETI): Formación, investigación y profesión. Madrid: AIETI.
- Comunicación del Consejo Educación y Formación 2010 (2004). DO C 104/01, de 30 de abril de 2004.
- Diaz-Aguado, M. J. (2003). Educación intercultural y aprendizaje cooperativo. Madrid: Pirámide.
- EF EPI. EF English Profiency Index (2015). http://media.ef.com/__/~/media/centralefcom/epi/downloads/full-reports/v5/ef-epi-2015-english.pdf
- European Commission (2014). The Erasmus Impact Study. http://ec.europa.eu/education/library/study/2014/erasmus-impact_en.pdf.
- European Commission/EACEA/Eurydice, 2015. The European Higher Education Area in 2015: Bologna Process Implementation Report. Luxembourg: Publications Office of the European Union.
- González, J. and Wagenaar, R. (Eds.) (2003). Tuning educational structures in Europe. Final report, phase one. Bilbao: Universidad de Deusto.
- Gros Salvat, B. (1995). Teorías cognitivas de enseñanza y aprendizaje. Barcelona: EUB.
- Huertas Barrios, E. (2013). *La competencia interpersonal en la formación de traductores en España: un estudio empírico-descriptivo del trabajo colaborativo durante la primera etapa de formación en Traducción e Interpretación.* PhD dissertation. University of Granada.
- Leuven and Louvain-la-Neuve Communiqué (2009). http://www.ond.vlaanderen.be/hogeronderwijs/bologna/conference/documents/leuven_louvain-la-neuve_communiqu%C3%A9_april_2009.pdf
- London Communiqué (2007). Towards the European Higher Education Area: responding to challenges in a globalised world. http://www.ehea.info/Uploads/Declarations/London_Communique18May2007.pdf
- Madrid Izquierdo, J. M. (2007). La política educativa de la Unión Europea al servicio del desarrollo económico con cohesión social. Revista Española de Educación Comparada, 13, 253-284.
- Marías, J. (2015). Ni bilingüe ni enseñanza. El País Semanal. http://elpais.com/elpais/2015/05/13/eps/1431541076_553813.html
- Ministerio de Educación (2010). Programa Integral de Aprendizaje de Lenguas Extranjeras. http://www.mecd.gob.es/prensa-mecd/en/actualidad/2010/10/idiomas.html
- Pérez Arias, Y. (2015). "Procesos de integración académica en Europa y Latinoamérica: Unión Europea, MERCOSUR y Alianza del Pacífico." Master's dissertation. University of Granada.
- Resolución del Consejo sobre «El desarrollo del capital humano para la cohesión social y la competitividad en la sociedad del conocimiento», (2003/C 295/05), (DO C 5.12.2003).
- Tymoczko, M. (2005). Trajectories of Research in Translation Studies. Meta, 50, 4. 1082-1097.
- Yorke, M. and Knight, P. (2006). Embedding Employability in the Curriculum. York: Higher Education Academy, Learning and Employability.

Humanities: History, Philosophy, Linguistics, Arts, and Journalism

What is the Role of Education in Plato's City State, and Why Does It Lead to Censorship? Is the Censorship in Plato's City State Justified?

Alexandra White
Student, University of Hertforshire, United Kingdom

1.INTRODUCTION

This essay will explain the different roles education has in Plato's city state, and also why most of them lead to censorship. The first and second paragraph explain why education is vital to Plato's city state, and this is because it teaches the guardian's the type of nature they should have and also keeps their souls healthy. The third and fourth paragraph are concerned with the negative impact of misrepresentation of heroes and gods and the reason why filtering what stories are told will lead to censorship. Another point I go on to make is about the physical training of the guardians mixed with them reading poetry and listening to music, and why I do not think it is justified to put restraints on how much one can train or read. My last paragraph will discuss the impacts of sexual intercourse, overeating and binge drinking on the guardians and why it must be censored in order to help them fulfil their duties to the city state. I will agree that the censoring of overeating and binge drinking is justified; however I will disagree with the censoring of sexual intercourse. Eventually, I will conclude that the overall censorship of the city state is unjustified as it does not allow people to express their individuality and treats everyone like they are machines. I will criticise Plato for not giving us a full picture of his society, as he seems to be focused on the role of education amongst the guardians, but neglects to tell us about the education of other members of his city state, such as the workers.

II. THE ROLE OF EDUCATION IN PLATO'S CITY STATE

The role of education in Plato's city state is to allow the guardians to develop the correct balance between being gentle and being tough. Plato states that the city state cannot rely on people's own natures, so they must be educated so that they can develop all of the right qualities to be just.

Plato locates justice on a wide scale, in his city state, before he locates it in the individual, as according to him, it is easier to comprehend justice on a bigger scale. Justice is maintained by the individuals who inhabit the city doing the roles that they are best suited for, for example the farmers should farm and the guardians should guard. The people in the city state should stay doing what they are inclined to do and all will be well. The role of education in Plato's city state is not only to maintain this justice, but also education is seen as a positive thing as it keeps the inhabitants' souls healthy. Without education, the people in the city state would have unhealthy souls, as they would not have been able to develop wisdom which is one of the key virtues to keeping the soul just. They would not have been able to develop wisdom, as wisdom is good judgement in your reason, and without education, they would be unable to make just decisions and therefore they would not have good judgement.

Education leading to censorship comes into this when Plato tries to filter what the guardians will learn, in order to make them just people. An example of the guardians' learning being filtered comes from this quote, from the Republic, *'our first business is to supervise the production of stories, and choose only those we think suitable, and reject the rest. We shall persuade mothers and nurses to tell our chosen stories to their children, and by means of them to mould their minds and characters'* (377b–c).[1] He uses the example of stories about heroes and God and the misrepresentation of both of their natures and he states *'contrary to the poets' claims, gods/heroes did not do harm, change shape, lie'* (377 e).[2] I would say this is pretty obviously because he does not want the young guardians to look up to heroes or gods and think that it is acceptable to do harm or lie because their role models do. However, this does lead to censorship as it means that the people of the city state will not be able to repeat certain stories, which they may want to tell. Proof of this is in the Republic when Socrates states his issues with the tale of Cronos, the states, such tales *'shall not be repeated in our state'* (378b).[3] In this specific situation, I believe that Plato is justified for censoring negative portrayals of the God's as it would be bad to have young guardians look up to bad role models if they are meant to be guarding the state. It could lead to them learning to do bad things, such as lying or stealing. He is not censoring stories about heroes and gods in general, he

[1] Plato, G. R. F Ferrari and Tom Griffith. The Republic. Cambridge. Cambridge University Press, 2000. Print. (377b-c)
[2] Plato, G. R. F Ferrari and Tom Griffith. The Republic. Cambridge. Cambridge University Press, 2000. Print. (377e)
[3] Plato, G. R. F Ferrari and Tom Griffith. The Republic. Cambridge. Cambridge University Press, 2000. Print. (377b)

just wants the ones which are told to be stories which represent them as good, he states, *'of our laws laying down the principles which those who write or speak about the gods must follow, one would be this: God is the cause, not of all things, but only of good'* (380c)[4]. But alternatively, censoring the stories that people can tell, and thus the stories that people believe, turns the people of the city state into robots who all think alike and a variation of opinions is not necessarily a bad thing. Censoring this will not allow people to form their own views as their views will be being formed through a filter of what Plato wants them to believe.

Plato did not want the guardians to fear death because they might be in situations where they are needed to go to war against another state, so they need to be brave. He also did not want them to fear death, as if they feared death they may feel pity towards their enemies after killing them, or alternatively, if a guardian from Plato's city state dies, the other guardians would need to remain focused, thus they should not be mourning. As a result of all of this, he did not want heroes and gods to be represented as fearing death and he also did not want Hades, the underworld, to be misrepresented as a bad place which the people in the city state should fear. A quote from the Republic which shows this is, *'we shall prevent (mothers and nurses) blaspheming the gods and making cowards of their children'* (381e).[5] This leads to censorship, as again, it puts a restraint on the stories people can tell, as it might lead to misrepresentations. In this situation, I do not think Plato is justified in making the guardians immunized to a fear of death. I think this because I believe people should be able to mourn loved ones passing. I do not think it would be much different in Plato's city state, as death is not necessarily represented as bad here – some people believe you get a promise of an afterlife and they are still upset at the passing of a loved one.
Another role of education, which I have discussed briefly is that it allows people to be fit for their assigned social roles, Plato states; *'a purpose of education is to create a balance, a harmonious state where the workers are to be trained to obey their masters and offer important economic services to the state.'* It prepares the philosophers to rule, as it gives them a better understanding of the difference between the intelligible world and the visible world, thus the knowledge to decipher appearance and reality. It also allows for the physical training of the guardians. The physical training of the guardians must be balanced with them reading poetry and listening to music, as if they engage in too much physical training, they will become too aggressive, whereas if they listen to too much music and read too much poetry, they will become soft. This again, in a way, leads to more censorship. They would be limited on how much poetry they are allowed to read and how much music they are allowed to listen to. He also states that certain music styles will have a negative impact on the guardian's character, so certain types of music need to be censored as well. I do not think this is justified as listening to music and reading poetry does not make someone too soft, it will just make them more knowledgeable of that which they are reading or listening to. Similarly, training will just lead to them becoming better at whatever it is they are practicing. I believe that Socrates goes too far in this respect, as if you start putting restraints on what people can do or what they can listen to, you are essentially taking away their basic human rights and controlling them. With the other censorships I have previously discussed, such as the misrepresentation of the gods, I can see why he would want to put restraints on bad stories being told, however, I believe this restraint is unnecessary and unjustified.

The final point that I will go on to discuss is the censorship of sex as a way of practising self-control. Plato also uses the examples of over-eating and binge drinking as forms of a lack of self-control. This does not really overlap with the role of education that much, but it could be argued that if you are taught the correct virtues, you will have a good judgement and will just naturally understand why giving in to temptation can promote a lack of self-restraint. However, Plato's mouthpiece, Socrates states that the topic of erotic love would be acceptable in a discussion of education, so in this respect, it fits into education, but maybe not so much the role of education.

Plato saw sexual intercourse as necessary for producing the next generation, but not necessary in any other way as it serves no useful end, other than giving the couple engaging in sex pleasure. If this is the case, it would also mean that homosexual relationships would be forbidden, as they only serve to pleasure, as you cannot procreate if you are homosexual. The aims you set out to fulfil are key to the health of your soul, and in this respect, being homosexual is not an aim that you can fulfil anything with, due to the fact you would be unable to procreate. This would be censoring when and if the people of the city can have sexual intercourse, it would be only when they want to procreate or if they can procreate. I do not agree with the censoring of this as it would prevent people from making deeper connections and also be removing people's right to choose how they want to

[4] Plato, G. R. F Ferrari and Tom Griffith. The Republic. Cambridge. Cambridge University Press, 2000. Print. (380c)
[5] Plato, G. R. F Ferrari and Tom Griffith. The Republic. Cambridge. Cambridge University Press, 2000. Print. (381e)

live their life and it would also go against one of the basic human rights - that you should be able to make your own decisions on the way you want to live your life. Although I disagree with this censorship, I can see why Plato would want to censor this, as if people fall in love or are seduced, it may lead to them being distracted, particularly if it is when the guardians are at war, it may lead to them being unable to guard the city state. Also, it would probably be beneficial if the guardians were not driven by their sexual desires and instead, practised having self-control. I do not think the censorship of overeating or binge drinking is a bad thing as it is in the interest of the guardians to have self-control when consuming these things. If the guardians overeat, they will become unfit, thus they will be unable to fulfil their duty to the city state. If the guardians binge drink, it can lead to ill health, and also can lead to them having bad judgement, as you do not think clearly when you are under the influence of alcohol. Censoring overeating and binge drinking would be acceptable as these two things have more of the capacity to be detrimental to the guardians career, however I do not agree with the censoring of sexual intercourse. The main reason that I do not believe this censorship is justified is purely because of the fact it is very unlikely that men will be seduced at war, thus the need for this censorship is essentially useless and it only takes away a pleasure from people, with no real ends.

In conclusion, I believe that overall, the censorship in Plato's state is unjustified. This is because, regardless of whether or not there are censorings on certain materials, people will still go against what rules are set out for them in order to feel like they have freedom. In Plato's city state, the people are all controlled like machines and what they are allowed to learn is at the hands of the rulers, thus I do not think they will lead happy lives. They will not be able to express their individuality through certain types of music or poetry, which I also think is a negative thing as it is good to have a mixture of people in society. Also, Plato seems very concerned with the role of education amongst the guardians, but seems to neglect to give us a picture of the workers and other members of his society. Thus, I feel like he neglects to give us a full picture of the city state and the role of education for those who are not guardians. To elaborate, the workers will probably not have to take part in some form of military training to gain strength, so will certain music be censored for them also? It is not clear how all of this censorship will impact society as a whole, but as it stands, I do not think the concept of a fever city could ever be put into place and work.

REFERENCES

- Ferrari, G. R. F. and Griffith, T. (2000). The Republic. Plato. Cambridge. Cambridge University Press, Print.

The Lack of European Awareness Regarding the African Continent The Case of Spanish Universities

José Manuel Maroto Blanco
Student, University of Granada (UGR), Spain

ABSTRACT

My objective is to discuss the issue of the lack of European awareness regarding the African continent, the birthplace of humanity, and the ethnocentricity which dominates the field of European academia. This attitude reflects the high self-interest of European society not only at home, but also in those places where it has extended its power. To this end, I will analyze and revise the history courses offered at Spanish universities in order to see precisely which ones offer education in the history of other continents in general and particularly in the case of Africa. Essential to this study is the knowledge of whether these courses are optional or obligatory. Connecting this academic context with European colonial history could well prove key to helping deconstruct false notions currently held about Western history being representative of world history.

1. EUROCENTRISM, IGNORANCE AND THE IMAGINING OF THE OTHER

Ethnocentrism entails a belief that the parameters and values defining a particular culture are qualitatively superior to any other. All cultures, in one way or another, tend both to exert some influence over others and to attempt to perpetuate themselves, although it would be unfair to argue that they all do this in the same way. If, as Silverstone (2010) claims, it is true that no identity can exist in the absence of a series of oppositions, it is also true that the historical events and processes that have defined Europe to date help to explain why Eurocentrism should have become one of the most exclusive systems of thought in the world. One such event in particular is undoubtedly the colonization of the Americas, for it is particularly at this point in time that European experience and outlook become transformed into a global vision in which not only are indigenous peoples considered primitive, ignorant and inferior, but so too is their way of looking at the world (Dussel 1992). Indeed, the concept of 'discovery' soon triumphs over that of 'conquest' in the conceptual mappings of European language, thus denying indigenous peoples, 'the other', any vestige of protagonism as shapers of human history.

Academically speaking, Eurocentric thought has been a manifest part, one way or another, of both the social sciences and the humanities throughout modernity and the history of contemporary Europe. Examples may be found in the explicit racism of certain strands of physical anthropology, which attempted to justify Western exploitation in the world by claiming a physical hierarchy between human groups – whether through the measuring of skulls or through the theorising of the origin of foreign cultures' mythologies. Both history and philosophy are also illustrative in this regard. Hegel's *Lectures on the Philosophy of World History* of 1837 (Hegel 1837) is an obvious example. Without a second thought, Hegel posits that the experiences of Africa and America lie outside history. Specifically in the case of Africa, Hegel unhesitatingly asserts a lack of the continent's historical interest, the complete absence of any technological contribution, and the barbarism of the way of life there. Such an ideology – while now possibly obsolete – is still clearly reflected in basic notions such as a division of Ancient, Medieval, Modern and Contemporary History which fails to take into account the timelines of Asia, Africa or Latin America: at the very least a sign of the minimal level of empathy towards other modes of viewing the world and a reflection of the hegemonic character of Western discourse.

As Edward Said (1996) has stated, this cultural battle by the West, one which employs notions and images of the 'other' rather than soldiers and canons, is also key to understanding Western culture. In this context, rationalism is seen as European heritage, the product of a system of thought which places Europe at the centre of World History, implying a definition of the periphery which in turn explains the very definition of European and Western culture itself (Dussel 2001). Such self-definitions retain at times a clearly racist character. Africa, for example, is not only considered part of this periphery; the sense of otherness is racialized, and *'is now both African and 'Negro', the latter being a necessary oppositional stereotype for the con- structing of white identities'* (Henry, p.139)

In terms of Western portrayals of Africa, I would like to make reference to the work of Antoni Castel, *Malas noticias de África* (2008). Castel, by compiling the conclusions of different authors' writings from the second half of the 20th century, shows how the negative image projected onto Africa in the West is also one which concerns our own definition of self. Here, Africa is presented as a place of disasters, of economic helplessness, of tribal conflicts based on primitive, irrational hatreds: it is a land in the thrall of religions and dictatorships. The underlying idea, however, is nothing less than that of a stable and peaceful West, composed of diverse and complex societies with democratic rule and popular sovereignty, whose religion and the state are separated and whose economic power allows for charitable intervention in Africa's problems. As such, Western violence, xenophobia and racism are hidden from sight together with the ever-widening gap in inequality, our shameful involvement in local conflicts, our unequal commercial relations with Africa, or the use of programs such as *Official Development Assistance* (ODA) as political tools in Western foreign policy.

2. THE INVISIBILIZATION OF AFRICAN HISTORY – A BRIEF LOOK AT THE SITUATION IN SPAIN

These portrayals of Africa in the West are not necessarily ones that have been clearly thought out; rather they have undergone continual distortion throughout our history. Yet the result is that Africa is nowadays seen as *'helpless, a fugitive in search of economic help and protection [...], either from Europe or from the U.S.'* (Iniesta 2006, p.11). In order to understand this, we need to look back at the history of relations between Europe and Africa. For Cheikh Anta Diop (2012), for example, recognising the links between Europe's technical superiority and its economic need for the exploitation and enslavement of black Africans is key to understanding the Western need to distort images of African abilities and morals. Such a contemptuous relationship, however, in no way impeded the commandeering of African resources, an appropriation which continues to this day. This is a relationship characterised both by ignorance of and prejudice towards blacks which dates back to ancient times and which reached the peak of its cynicism in the idea of 'the white man's duty', the perceived need to colonise Africa and educate its people to the same cultural standards as elsewhere. Here the double standard is evident, as Western interests are masked by a call to moral decency and responsibility; something very evident in the verses of Rudyard Kipling's poem, *The White Man's Burden* (1899). The poem, while it refers specifically to the situation of the North Americans in the Philippines, clearly reflects the way of thinking of a large part of Western society:

Take up the White Man's burden,
The savage wars of peace -
Fill full the mouth of Famine
And bid the sickness cease;
And when your goal is nearest
The end for others sought,
Watch sloth and heathen Folly
Bring all your hopes to nought.

As contact between the West and Africa became closer, such stigmas became more and more present in Western discourse. Through an association of the concepts *African* and *savage*, Africa was portrayed in a supposed state of cultural paralysis. Indeed, not only was being black associated with emotion and distanced from reason; until recently most schools of anthropology still saw non-Western cultures from the point of view of a one-way evolutionary process, with European society placed at the summit. A case in point is the general European view of Egypt, in which Pharaohs are seen as white and slaves as black. Champolion, for example, denied the possibility of the first Egyptians being black (Iniesta 1997). For Diop (2012), any such denial also implies the denial of the influence of black cultures on the birth of Greece, the mother of Western civilization. In contrast, Diop has argued that the Ancient Egyptian practises of totemism and circumcision, a vitalist conception of royalty ¬– very much in common with the rest of Africa – together with the similarities between Egypt and other African cultures in cosmogony, social organization and the importance given to matriarchy, all typify Ancient Egypt as a black African culture.

Figure 1,2. Members of Egyptian Royalty

From left to right: the Pharaohs Narmer and Mentunhtep II, whose black facial features are evident.

As Iniesta points out (2007), this traditional perspective goes as far as denying there can be *any reading* of African history whatsoever, as prior to European colonisation no history of this kind supposedly existed. Africa, then, is still held captive by the Hegelian claims mentioned above: it continues to be relegated to a non-historical status by a European thought full of ignorance, disinterest and prejudice. Proof of such a claim can be seen in the type of elitist university education given during the colonial era, one which ignored African history completely. In the case of Spanish colonial education in Equatorial Guinea, the main African quality presented was initially that of *savagery*. This in time gave way to another, related concept: the idea of black Africans as *homo infantilis*, which in turn placed more emphasis on the idea of an Africa whose people were in need of protection. Here again, features of Hegelian thought can be seen. Such propositions would become an argument with which to justify Spanish control and the colony's submission (Fernández-Fígares 2003, p. 18; Bandrés & Llavona 2010). In a similar vein, attempts were made to impose Spanish gender roles throughout Guinean youth society, as these were considered to be more dignified than local practices.

This Western conception of African savagery can be partially explained by reports of native populations made by European explorers. Manuel Iradier, the first Spaniard to write about the Fang of Equatorial Guinea, provides us with an exotic image of Africa in his writings in which cannibalism appears as a very real threat. This desire to attribute savage behaviour to African populations, however, was also common in the rest of Europe and and stemmed from a longstanding medieval traditional later reinforced by European expansion in Africa. *Mission to Cape Coast Castle and Ashante* (1819) by Edward Thomas Bowdich, *Voyage au Como* (1861) by Braouezec or *Explorations and Adventures in Equatorial Africa* (1861) by Paul Belloni Du Chaillu are just some examples which paint a similarly savage image of African populations (Sánchez 2011).

Indeed, both literature and the press have played a pivotal role in the formation of the European image of Africa. Back in 1914, the writer José Más, author of novels such as *En el país de los bubis* (1919 - inspired by Fernando Pó, now the island of Bioko in Equatorial Guinea) strikingly denounced Spain's lack of knowledge of its Sub-Saharan colonies in the magazine *África Española*. Such was the level of ignorance that in the national press of the day one could, for example, read news reports that talked of problems with the colony's electrical wiring systems at a time when no cabling had even been installed in the area. The Spanish national press was also instrumental in pushing for campaigns in Africa as compensation for Spain's loss of its American colonies (Sánchez 2014). Newspapers would continue to feed on ideas and language drawn from a historically racist tradition to the extent that later on some would even manage to elegantly accuse Nigerian workers who were then working on the island in conditions of semi-slavery of 'politely staging a growing invasion of the island' (ABC, 22nd January, 1964, *'Guinea y sus vecinos. Nigeria, al otro lado del mar'*). Even today, there exists in the press a distorted image of Africa that finds its origins in the colonial tradition. The recent treatment of Nigeria in the Spanish press is a case in point. The fact that ideas taken from the colonial era (such as that of Africa as 'the white man's grave') are still alive today can be seen in the recent extensive coverage given to the Ebola virus – even though it causes infinitely fewer deaths than either AIDS or malaria. The relevance here lies not, in fact, in the seriousness of the outbreak but in the possibility of its transmission to the 'white man'. Similarly, the general image presented of Nigeria is still one of violence and disasters, which in many cases are portrayed as implicitly related to either Islam or negritude (Maroto 2015).

Against this continual backdrop of invisibilization and distorted realities, we Europeans continue to remain ignorant of the history of our neighbours, one which is nevertheless very much a part of our own. It is high time therefore that we recognise our involvement in the historical events which have taken place in areas of the world such as Africa. I believe it to be a good way of decolonising our own selves, something which is tremendously necessary and which, as Jean Paul Sartre stated, involves the deconstruction of our own ethnocentric discourse with the help of new perspectives from researchers from other traditions from other parts of the world. To this end, and following Sanou Mbaye (2010), I would like to make special mention of Sundiata Keita, the founder of the Mandiga Empire and the originator of the Manden Charter (Kurukan Fuga) of 1222. The charter, now considered the first declaration of human rights, puts into question the idea that Europeans were necessarily the first to institutionalise ideals of fairness. The following extract is a clear example of the recognition of individual liberty, the endeavour for solidarity, and the opposition of slavery:

'The hunters declare: all human life is one. [...] but no life is 'older', more respectable than another. [...] The hunters declare: hunger is not a good thing neither is slavery; they are the worst calamities that can occur en this lower world. [...] from now on, nobody will place the bit in the mouth of his equal to lead him to sale; neither will anybody be beaten, much less executed, because they are the child of a slave. The hunters declare: from today, the spirit of slavery has

been laid to rest [...] Therefore, the hunters declare: from now on, each one is master of their own person; from now on, each one may act freely, each one has a right to the fruits of their labour. This is the oath of Manden, let the whole world hear.'

(Mbaye 2010, pp.133-4)

3. IS THE CURRENT EDUCATION SYSTEM ETHNOCENTRIC? A STUDY OF THE TREATMENT OF NON-EUROPEAN QUESTIONS IN THE SPANISH UNIVERSITY SYSTEM

Many would argue that education in contemporary society is to a great extent an instrument at the service of the state. It is hardly surprising therefore that the content of school education has become little more than a reflection of what the politicians and rulers of each era have wanted to transmit to the coming generations. Evidently, the very concept of education has not remained the same throughout these periods, given that it has itself undergone changes according to the historical contexts in question and has gone from being from a privilege limited only to the rich to an inalienable right for all as part of the welfare state. As we have seen, ethnocentrism was previously a reality in Western education and, despite the intervening years, Eurocentrism continues to dominate in education even as far down as infant school. It is for this reason, therefore, that Europeans of Maghrebi, Asian, or Sub-Saharan African origin often end up feeling excluded when they study a European history which refuses to take into account the ethnic diversity which modern European countries are made up of.

Nevertheless, such Eurocentrism goes beyond the education provided in Western countries. A study by Albuquerque and Ibarra (2014) concluded that with the exception of Argentina, Bolivia and the Dominican Republic, history education in Latin American countries focuses more on the history of the United States and Europe than on the rest of Latin America. They also highlighted how in Ecuador and Venezuela, countries with a marked Pan-American political discourse, both local and international history was studied with a noticeably Eurocentric emphasis to the detriment of Latin American study. Here, it is also worth noting the deficiencies commented on by Vega (2007), who highlights the fact that in many cases the Latin American context is studied in both middle and high school from a purely European perspective. Against this, Vega proposes the development of other, less Eurocentric models than those currently in use in education system and in universities in particular.

In terms of African education, Paul Tiyambe Zelaza (2009) has noted the paradox which existed under colonialism in African universities, who did not even teach African history and relegated the entire continent to the field of anthropological study. Although African education initially received a strong boost after the former colonies' independence, it later suffered under the structural adjustment programs of the 1980s and 1990s as governments cut education budgets and began to focus mainly on primary schools. Nowadays, however, there is competition between a number of dominant cultural positions, among which the *globalist, deconstructionist, developmentalist* and *culturalist* all stand out. Each of these is different from the other and each contains a different level of Eurocentrism. While, for example, the culturalist perspective has led to a more Afrocentric focus, numerous studies and new disciplines, the globalist has conserved hugely Eurocentric roots which ¬–obliging Zelaza (2009, p.131) himself to ask the question, *'can African studies escape –even transcend– the Eurocentric coding, the seductions and sanctions of writing Africa by analogy?'*

Neither has criticism of the strongly ethnocentric element of university studies in Europe been lacking. In Introduction: *From University to Pluriversity: A Decolonial Approach to the Present Crisis of Western Universities,* Boidin, Cohen and Grosfoguel (2012) present a series of studies which clearly show the urgent need for European universities to reinvent themselves: to become capable of both adapting themselves to and of making use of an intercultural dialogue which might bring about a Pluriversity. The book criticises a number of different European universities. Among the most notable are papers criticising the university situation in Holland, where 'minority research' has been dominated by a dominant elite which considers minorities groups purely from the point of view of population dynamics (Nimako 2012), and where certain historians have tried to minimize Dutch involvement in the slave trade (Hira 2012). Another interesting paper by María Paula Meneses (2012) criticises, on the one hand, a Portuguese historiography which has tried to silence Portugal's colonial involvement and, on the other, a Mozambican historiography whose ambitions of creating a nation state have led to the silencing of the many and different historical accounts of the country's colonial experience.

In the case of the Spanish university and education system, there has been no real questioning of the Eurocentric model currently being followed. The criticisms and improvements proposed both in the university system as well as the general education system at large have focused on problems brought about as a result of a lack of a centralised state education policy: the more than eight educational reform bills which have taken place in the last forty years; the controversial entry of private business in the state university system; and the need for a perspective which would take into account both gender and (in the case of historical studies) the multiple perspectives of state and regional political history, an issue which generates controversy to this day.

A cursory examination of current Spanish government legislation regulating the compulsory curricula of secondary and high school education (Edict thirty-seven of the BOE, passed on 03/01/2015) shows that non-European studies are practically non-existent. In fact, African issues are only studied as a brief part of one subject in final-year secondary studies entitled 'the process of decolonisation in Asia and Africa' which completely ignores the role of Spain in the colonisation of Sub-Saharan Africa. This silence amounts to the erasure of more than 200 years of Spanish exploitation in the current territory of Equatorial Guinea at a single stroke. The problem is further exacerbated by a continued lack of familiarity with non-European issues at university level, despite the fact that universities supposedly enjoy greater autonomy in their choice of courses than do secondary schools. Indeed, the limited knowledge of both Africa and Asia made available in Spanish universities is an undeniable reality. An analysis of the courses currently on offer gives serious pause for thought over just how knowledge is built in the university system, both in Europe and in Spain in particular.

It is clear that both African and Asian studies form part of the periphery of university courses, to the extent that in some cases both continents form part of a single subject. Furthermore, Asian studies often focus solely on one country, the Philippines, and on little more than on Spanish influence there.

Figure 3. Non-European Courses Offered in Spanish Universities by Province

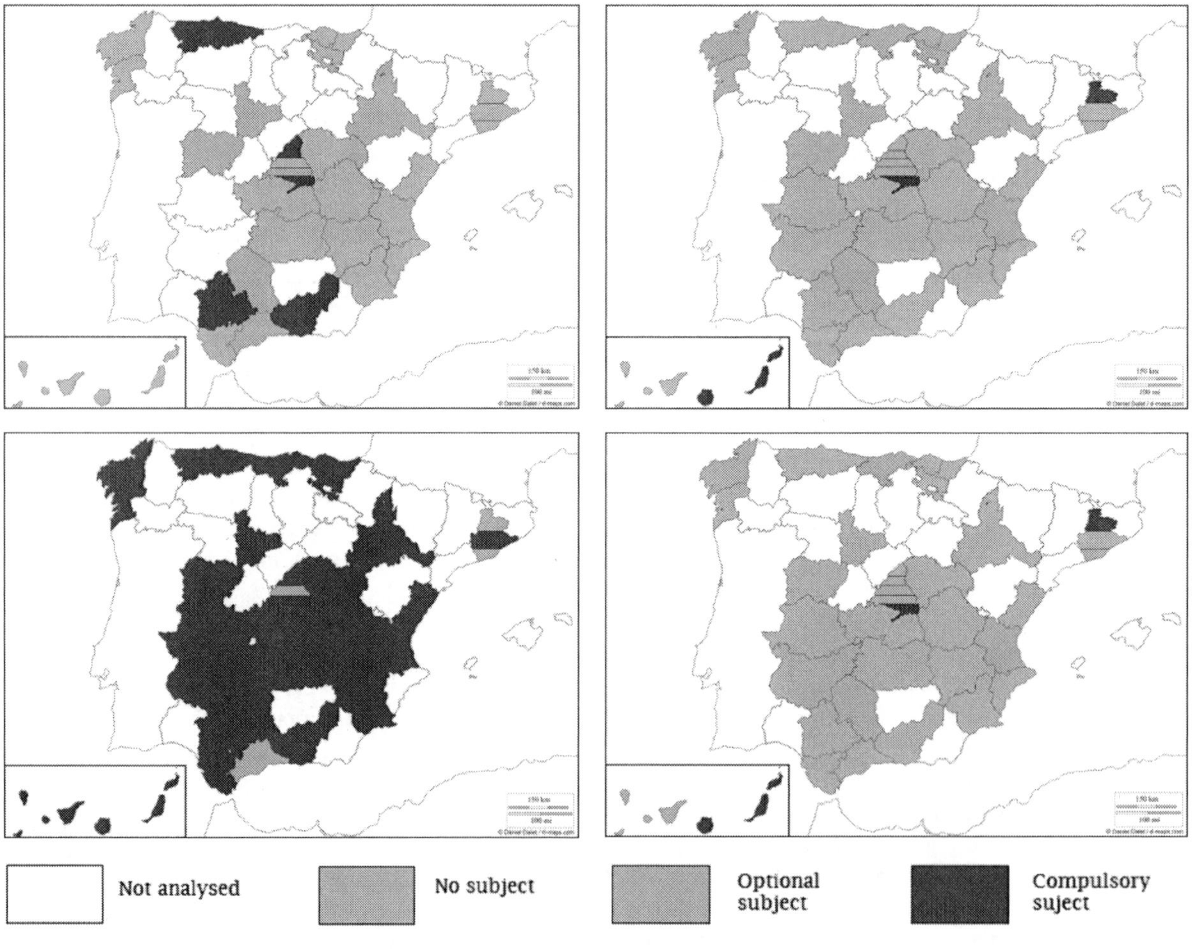

Middle East (above left), Asia (above right), Latin America (below left) and Africa (below right).

Incredibly, there is not a single course which focuses on the period of Spanish colonisation in Africa, much less one which might deal with the period before that, in which these lands were exploited by the Spanish upper classes.

Elsewhere, a very different phenomenon can be seen. Latin American studies certainly epitomise the idea that the history of Latin America is of interest only insofar as Europeans (in this case the Spanish) settled there. There is not a single Spanish university which does not offer courses on Central and South America; in fact, they are compulsory (see Figure 3). Nevertheless, the periods studied correspond only to those in which the Spanish were present, and clearly demonstrates that the academic interest herein lies in the importance of Spanish involvement. Indeed, it is a striking sign of academic priorities that courses on pre-Columbian periods should be so much scarcer and merely optional in nature. Or that in the few courses where a reference is in fact made to pre-Columbian peoples (such as the History of modern America course) this introduction is clearly designed to explain their political infrastructure as a precursor to understanding the process of conquest. Such Eurocentrism, however, is more complex than mere indifference towards 'the other', or a characterisation of their importance simply in terms of a Spanish presence. In the case of those courses which specifically focus on the history of the Middle East, what is interesting is that all make reference to ancient history, and all clearly state two fundamental ideas: 1) the technical superiority of the Middle East with regard to Europe at that time (for which reason it is considered a subject worthy of study); 2) a teleological vision of history, an idea with clearly Western roots in which advances are made in linear form towards progress. Nor is it strange to find a course on the history of Islam, yet this is always in relation to the presence of Islam in Spain.

4. CONCLUSIONS AND REFLECTIONS

The approach of European universities to the knowledge of non-European situations currently remains an extremely unresolved issue. While it is true that each country has approached the subject differently according to its particular historical context, it is also true that Eurocentrism is in force far beyond traditional Western borders, in parts of the world such as Latin America or Africa. In the case of Spain, two clearly different attitudes are evident. On the one hand, there exists an attitude of complete indifference and ignorance towards those situations considered to be alien, such as pre-Columbian America, Asia, the Middle East or Africa. On the other, there is enormous interest in Latin America when the object of study is what is termed *Modern and Contemporary History* – a clear demonstration that academic interest in the continent is directly linked to Spain's physical or economic presence there. A third concern is the minimal, not to say zero, academic interest in the Spanish colonisation of Sub-Saharan Africa and the Maghreb. There is a sense here that history is being 'racialized', where the absence of light-skinned colonists in Morocco, the Sahara, or Equatorial Guinea seems more than motive enough to ignore a part of history which affects us directly. Here again, the whitewashing of yet another period of colonial exploitation apart from the Americas may well be a strong argument for avoiding the issue.

In Spain we continue to justify Hegelian ideas of other continents' lack of history to the extent that the Spanish university system seems a long way indeed from its purported emancipatory ideals – and further still from any model of *Pluriversity.* Against such a background it seems a matter of urgency that we ask ourselves whether it is possible to build the common, global projects which will be so necessary to our future if we do not accept that the peoples we speak to have structures of thought and epistemologies different from the West's, fruit of their own rich and diverse histories. We need to ask ourselves whether denying ourselves the historical knowledge of these cultures is indeed beneficial, or whether on the contrary, depriving ourselves of such information is not in fact an act of cultural suicide. It is time to recognise the diverse origins of the neighbours who make up part of our cities, to learn of their histories and of their cultures. Such a deepening of understanding can only help improve both our social relations and our sense of community. In this regard, it is a key that educational institutions start to give greater prominence to diversity. As I write, thousands of refugees are arriving in Europe from Syria, risking their lives because 'death comes but once'. In Europe we refuse to take them in, while other countries closer to Syria are caring for them with far fewer resources available. If we now fail to understand the history of our neighbours, if we are incapable of understanding their part in our own, if we are no longer able to empathize with the most destitute among them – then, to our great regret, the humanity of the West will eventually become something studied only in history books.

REFERENCES

- ABC (2 January, 1964). Guinea y sus vecinos. Nigeria, al otro lado del mar, 27. http://hemeroteca.abc.es/nav/Navigate.exe/hemeroteca/sevilla/abc.sevilla/1964/01/02/027.html.
- Albuquerque, G. and Ibarra, P. (2014). La enseñanza de la historia en Chile y América Latina: análisis comparativo de contenidos curriculares. Educación media y equivalentes. Santiago de Chile: Universidad Bernardo O'Higgins. Vicerrectorado de Desarrollo. Dirección de Investigación. Observatorio Regional de Paz y Seguridad. Centro de Estudios Históricos. http://www.orpas.cl/wp-content/uploads/2014/09/Estudio1.pdf.
- Bandrés, J. and Llavona, R. (2010). Psicología y Colonialismo en España (II): en busca del Cociente Intelectual del Negro. *Psychologia Latina, 1(2),* 154-162.
- Boidin, C., Cohen, J., and Grosfoguel, R. (2012). Introduction: From University to pluriversity: A decolonial approach to the present crisis of Western universities. Human Architecture. *Journal of the Sociology of Self-Knowledge, 10(1),* 1-6.
- Castel, A. (2007). Malas noticias de África. Barcelona: Bellaterra.
- Diop, C. A. (2012). Naciones negras y cultura. De la antigüedad negroegipcia a los problemas culturales del África Negra hoy. Barcelona: Edicions Bellaterra.
- Dussel, E. (1992). El encubrimiento del otro. Hacia el origen del mito de la modernidad. Madrid: Nueva Utopía.
- Dussel, E. (2001). Eurocentrismo y modernidad (Introducción a las lecturas de Frankfurt). In W. Mignolo (Ed.): Capitalismo y geopolítica del conocimiento. El eurocentrismo y la filosofía de la liberación en el debate intelectual contemporáneo. Buenos Aires: Ediciones del Signo.
- Fernández-Figares Romero de la Cruz, Mª. D. (2003). La colonización del imaginario. Imágenes de África. Granada: Editorial Universidad de Granada.
- Hegel, G.W.F. (1974). Lecciones sobre filosofía de la historia universal de Hegel. Madrid: Revista de Occidente.
- Henry, P. (2004). Between Hume and Cugoano: Race, Ethnicity and Philosophical Entrapment. The *Journal of Speculative Philosophy, 18(2),* 129-148
- Hira, S. (2012). Decolonizing the mind: the case of the Netherlands. Human Architecture. *Journal of the Sociology of Self-Knowledge, 10(1),* 53-68.
- Iniesta, F. (1997). El planeta negro. Aproximación histórica a las culturas africanas. Madrid: Los libros de la catarata.
- Iniesta, F. (2006) El estigma de Cam. El negro en el pensamiento occidental. In A. Castel y J. C. Sendín (Ed.): Imaginar África. Los estereotipos occidentales sobre África y los africanos. Madrid: La Catarata.
- Iniesta, F. (2007) Kuma. Historia del África negra. Barcelona: Edicions Bellaterra.
- Nimako, K. (2012). About Them, But Without Them: Race and Ethnic Relations Studies in Dutch Universities. Human Architecture. *Journal of the Sociology of Self-Knowledge, 10(1),* 45-52.
- Mas, J. (30 August, 1914). De la Guinea Española, Verdades Amargas. Revista: África Española: revista de colonización, industria, comercio, intereses morales y materiales, 422-424.
- Maroto Blanco, J.M. (2015). El infoentretenimiento en la cobertura mediática sobre Nigeria: Boko Haram y el ébola como ejemplos del etnocentrismo y de la banalidad de la realidad. In: Mateos, C. y Herrero, J. (Ed.) La pantalla insomne. San Cristóbal de La Laguna: Cuadernos Artesanos de Comunicación, 372-413.
- Mbaye S. (2010). África al socorro de África. Madrid: Los Libros de La Catarata.
- Meneses, M. P. (2012). Images Outside the Mirror?: Mozambique and Portugal in World History. Human Architecture. *Journal of the Sociology of Self-Knowledge, 10(1),* 121-136.
- Pérez J. and Vega V. (2007). La enseñanza de la historia contemporánea de América Latina en las universidades del Cono Sur, Rosario: Prohistoria Ediciones.
- Said, E. W. (1996). Cultura e imperialismo. Barcelona: Editorial Anagrama.Sánchez Molina, R. (2011). El pamue imaginado. Los fang de la literatura colonial española. Navarra: Editorial Aranzadi.
- Silverstone, R. (2010). La moral de los medios de comunicación. Sobre el nacimiento de la polis de los medios. Buenos Aires: Amorrortu.
- Zeleza, P. T. (2009). African Studies and Universities since Independence: the challenges of epistemic and institutional decolonization. *Transition, 101(1),* 110-135.

G200 Association
14 rue du Rhone
Geneva 1204, Switzerland

CURRAN ASSOCIATES INC.
proceedings.com

ISBN 978-1-5108-3254-1

9 781510 832541

6th International Conference on Chemistry and Chemical Process (ICCP 2016)

MATEC Web of Conferences Volume 49 (2016)

Rome, Italy
1 - 3 February 2016

edp sciences